COEVOLUTION OF FUNGI WITH PLANTS AND ANIMALS

COEVOLUTION OF FUNGI WITH PLANTS AND ANIMALS

Edited by
K. A. PIROZYNSKI
Paleobiology Division, National Museum of Natural Sciences, Ottawa, Canada

D. L. HAWKSWORTH
CAB International Mycological Institute, Kew, Surrey, UK

1988

ACADEMIC PRESS
Harcourt Brace Jovanovich, Publishers
London San Diego New York Berkeley
Boston Sydney Tokyo Toronto

ACADEMIC PRESS LIMITED
24/28 Oval Road, London NW1 7DX

United States Edition Published by
ACADEMIC PRESS INC.
San Diego, CA 92101

British Library Cataloguing in Publication Data
Pirozynski, K. A.
Coevolution of fungi with plants and animals.
1. Animals & plants. Coevolution wtih fungi
I. Title II. Hawksworth, D. L.
(David Leslie), *1946–*
575
ISBN 0-12-557365-0

Typeset by Eta Services (Typesetters) Ltd, Beccles, Suffolk
and printed in Great Britain at The Alden Press, Oxford

Contributors

J. Bissett *Biosystematics Research Centre, Agriculture Canada, Ottawa, Ontario K1A OC6, Canada*

A. Borkent *Biosystematics Research Centre, Agriculture Canada, Ottawa, Ontario K1A OC6, Canada*

B. Boullard *La Boissière, 2346 rue de la Haie, 76230 Boisguillaume, France*

K. Clay *Department of Biology, Indiana University, Bloomington, Indiana 47405, USA*

M. W. Dick *Department of Botany, Plant Science Laboratories, University of Reading, Whiteknights, Reading RG6 2AS, UK*

H. C. Evans *CAB International Institute of Biological Control, Silwood Park, Ascot, Berkshire SL5 7PY, UK*

D. L. Hawksworth *CAB International Mycological Institute, Ferry Lane, Kew, Surrey TW9 3AF, UK*

T. Hijwegen *Department of Phytopathology, Agricultural University, P.O. Box 8025, 6700 EE Wageningen, The Netherlands*

D. W. Malloch *Department of Botany, University of Toronto, Toronto, Ontario M5S 1A1, Canada*

K. A. Pirozynski *Paleobiology Division, National Museum of Natural Sciences, P.O. Box 3443 (Station D), Ottawa, Ontario K1P 6P4, Canada*

D. T. Wicklow *Northern Regional Research Center, Agricultural Research Service, United States Department of Agriculture, 1815 North University Street, Peoria, Illinois 61604, USA*

To D.B.O. Savile in recognition of
his pioneering studies on the
coevolution of fungi and their hosts

Preface

Perhaps two-thirds of the known species of fungi form intimate relationships with other living organisms in parasitic, commensalistic, or mutualistic symbioses. The evolutionary implications for the fungi and hosts involved have been almost entirely ignored, except for certain pathogens of cultivated plants. The mainstream of coevolutionary debate during the last twenty years has tended to have little impact on the perspective of mycologists. It is not that coevolution has not been a major factor in the evolution of both fungi and their hosts, but rather that mycologists, with a few notable exceptions, particularly Dr D. B. O. Savile, have failed to appreciate its significance.

In our view, fungi have had and still exert a major influence on the evolution of their hosts, in turn adapting themselves to the changing host genotype to produce increasingly intimate coevolved situations. The object of this volume is to draw attention to the wide range of associations between fungi and living organisms, particularly animals and plants, where there are indications that coevolution has been a major factor in their development. In many instances there is no proof of the reciprocal genetic changes that are the hallmark of coevolution; rather there is circumstantial evidence inferred from ecological and biological observations. In order to emphasize the probable scale of coevolutionary effects involving fungi, we urged our contributors to be speculative and not to shrink from proposing novel hypotheses for future testing. Major advances in our understanding of such complex evolutionary situations are unlikely without hypotheses being exposed for critical debate and rigorous scientific scrutiny.

Following an overview, which includes a discussion on the concept of coevolution and draws attention to aspects not considered in depth later in the volume (Pirozynski and Hawksworth, Chapter 1), contributions are first grouped according to whether the fungi are associated with photosynthetic organisms. Dick (Chapter 2) analyses the situation in the heterokont fungi (Protoctista) and draws particular attention to the relationships between downy mildews and their host angiosperm families and biochemical considerations. The coevolution of pathogenic fungi with their hosts is reviewed by Hijwegen (Chapter 3), who emphasizes the problem of "jumps" to disparate hosts and the practical implications of these in crop protection and systematics. We have deliberately omitted chapters focusing on well-known "arms-race" situations involving plant pathogens because recent reviews of the coevolutionary significance of these have been provided by Drs J. A. Barrett and J. E. Parlevliet*. Clay (Chapter 4) draws attention to a probably widespread mutualistic association between fungi and vascular plants that is becoming increasingly recognized as of evolutionary importance—that of fungal endophytes producing toxins protecting the host from herbivores. Coevolution with symbiotic fungi in hepatics, a particularly ancient trait, is described by Boullard (Chapter 5) and is considered a significant factor in the evolution of liverworts even though the physiology of the relationship remains obscure. In the case of mutualistic symbioses between fungi and algae and (or) cyanobacteria forming lichens, Hawksworth (Chapter 6) describes the various coevolved structures, products, and strategies; the physiological and biochemical relationships are clearer in lichens but interesting questions arise regarding "altruism" in the photosynthetic partners.

The second group of contributions focuses on fungi associated with animals. In the case of entomogenous fungi, Evans (Chapter 7) finds that there is overwhelming circumstantial evidence for coevolution amongst entomopathogens and discusses the biology and phylogeny of these in relation to those of their hosts. Secondary metabolites and their role as chemical defence systems have particularly wide coevolutionary implications for both arthropods and rodents, and have practical relevance in bicocontrol; these aspects are considered in depth by Wicklow (Chapter 8). As a particular example of the intimacy that has coevolved between fungi and insects, we have selected that of ambrosia galls, in which gall midge larvae have specific fungal symbionts (Bisset and Borkent, Chapter 9); the association is apparently mutualistic, and cladistic studies indicate it to be plesiotypic in gall-forming cecids.

In the case of mycorrhizal plants, circumstantial evidence is provided by Pirozynski and Malloch (Chapter 10) for coevolved structures and strategies in the concomitant dispersal of spores and seeds by vertebrates, in grasses and forbs by large herbivores and small carnivores, and ectomycorrhizal forests with masting, edible seeds, hypogeous fungi, and fungus-eating vertebrates. Horizontal gene transfer between organisms of different kingdoms has been established in *Agrobacterium*. Pirozynski (Chapter 11), extrapolating from data on this bacterium and endophytic and gall-forming fungi, speculates that assimilation by the host plants of fungal genomic material able to induce deformations may have given rise to angiosperm organs such as flowers, fruits, and food-storage organs. If this is indeed even partly the case, the study of the coevolution between fungi and their hosts should come

* In A. R. Stone and D. L Hawksworth (eds), *Coevolution and Systematics* [Systematics Association Special Volume no. 32], Clarendon Press, Oxford, 1986.

to occupy a central position in discussions not merely of the evolution of fungi but of flowering plants as a whole.

We found the research carried out in preparing this book opened exciting new fields for reflection, hypothesis, and critical study in this hitherto neglected area of evolutionary biology. If it serves to stimulate enquiry and debate we shall consider its objective realized.

This book was conceived during the Third International Congress of Systematic and Evolutionary Biology (Brighton, UK; 5–10 July 1985), when we organized a symposium on this topic. However, the major part of this book was prepared while K.A.P. was on study leave at the CAB International Mycological Institute at Kew in 1986–87.

Ottawa and Kew
January 1988

K. A. Pirozynski
D. L. Hawksworth

1 Coevolution of Fungi with Plants and Animals: Introduction and overview

K. A. PIROZYNSKI

Paleobiology Division, National Museum of Natural Sciences, Ottawa, Canada

AND

D. L. HAWKSWORTH

CAB International Mycological Institute, Kew, Surrey, UK

Abstract

Fungi appear to be historically primed and biologically equipped to interact with other organisms on an intimate, stable, and protracted basis. Intimate symbioses, as exemplified by fungal parasitism of, and mutualism with, plants and animals, are most likely to provide opportunities for, are balanced by, or result from coevolution.

Little is known about progressive reciprocal genetic changes in co-symbionts involved in such associations. Following a synopsis of symbioses involving fungi, viruses, bacteria, protoctists, and other fungi, those between fungi and plants and animals that offer potential for a more critical study of coevolution are tentatively identified. Specific examples drawn from this book and elsewhere are arbitrarily categorized into symbioses between fungi and plants (e.g., plant parasites, mycorrhizal mutualists), fungi and animals (e.g., entomopathogens, mycangial mutualists of wood-eating arthropods), and those involving all three kinds of organisms (e.g., *Septobasidium*, vectors of plant parasitic fungi, ambrosia galls, fruit-flies and yeasts, moulds of rodent seed caches, endophytes and herbivores, myrmecophiles). The arrangement is convenient but simplistic, as "one-to-one" associations usually involve symbiotic systems rather than isolated organisms, such as mycorrhizal plants, and animals with symbiotic gut micro-organisms.

The relationships between individual components of a symbiosis are evaluated in terms of what is conventionally regarded as antagonistic or mutualistic. This distinction is also arbitrary, because ecological benefits and liabilities of symbioses have rarely been adequately evaluated from the coevolutionary perspective and may not be readily apparent.

COEVOLUTION OF FUNGI
WITH PLANTS AND ANIMALS
ISBN 0-12-557365-0

INTRODUCTION

In coining the term *coevolution* to describe reciprocal adaptations between plants and butterflies leading to their interdependence, Ehrlich and Raven (1964) opened literary floodgates; coevolutionary approaches became fashionable in almost all studies of plant–animal interactions. The timing was opportune. The Synthetic Theory of evolution, challenged by molecular and palaeontological data, came increasingly under fire from scientists who insisted that hypotheses must be falsifiable. "Natural selection ... can be used to explain everything and thus, unfortunately, it explains nothing" (Greenwood, 1981, p. 99). More significantly there was increasing realization that "the biotic environment, unlike the abiotic one, has the crucial property to evolve through its interaction with the population" (Law, 1985); that is, that genetic change may be "driven both by immediate interactions and by the feedback through the rest of the community" (Futuyma and Slatkin, 1983). Evolutionary adjustment to the latter is coevolution. Its reality rocked the foundations of theories of evolution, population genetics, and community ecology, each professing genetic sanctity of individual species, and put to question some of the more easily quantifiable demographic-versus-sociobiological approaches to population biology (e.g., Rabotnov, 1980).

At about the same time, Margulis (Sagan, 1967) published the first of a series of contributions that restated arguments for the Serial Endosymbiotic Theory of the origin of eukaryotic cells and contributed to a renewed interest in symbiosis (Ahmadjian and Paracer, 1986; Smith and Douglas, 1987) and its pertinence to coevolution. Symbiosis, defined in the original sense (de Bary, 1879) as the living together of different organisms, embraces most of life (Starr, 1975; Lewis, 1985). The term is now used to encompass interactions ranging from the integration of alien genes and their expression in another organism (Taylor, 1983; Verma and Nadler, 1984) to intermittent associations, as in pollination and seed dispersal mutualisms (Janzen, 1985). Coevolution can be interpreted equally broadly, as occurring at any point along the symbiotic continuum and beyond, between nucleotide sequences within a genome or between individual cells of a multicellular organism at one extreme and between biological and physicochemical elements on a global scale (Lovelock, 1986) at the other (Van Valen, 1983).

Within the span of the symbiotic continuum, coevolution can be equated with symbiosis in that "Nothing has evolved in isolation. . . . Coexistence can be considered a major consequence of coevolution when it describes the outcome of those evolutionary processes which enable different organisms to share the resources of a habitat" (Greenwood, 1981, p. 97). Defined as broadly as this, "coevolution encompasses much of evolution and can hardly lay claim to any distinction as a subject of study" (Futuyma and Slatkin,

1983). Even when restricted to a component of evolution focusing on the contact between evolution and ecology, and even if restricted to mean that "coevolution occurs when the direct or indirect interaction of two or more evolving units produces an evolutionary response in each" (Van Valen, 1983), coevolution may comprise most of evolution. The range of biological implications of coevolution has been discussed by Thompson (1982).

Not all adaptations are responses to interactions between living organisms, and Janzen (1980) warned that "coevolution" must not be used synonymously with "symbiosis" though it frequently is. Coevolution may be mimicked by "sequential adaptations from different causes or simultaneous adaptations to the same environment" (Van Valen, 1985). The majority of interactions, as those between vascular plants and animals on which much work on coevolution is based, are "diffuse"; they involve interactions between taxa of variable rank and asymmetric evolutionary rates (Herrera, 1985) or interactions in which "arrays of populations reciprocally affect each others' evolution" (Janzen, 1985), i.e. in which the characters being selected are influenced by arrays of interacting genes. "Almost all seed dispersal mutualisms are diffuse and non-obligatory for any one participant in that numerous species and individuals make up the disperser coterie of a given plant or plant population and they also depend on many other sources of food" (Janzen, 1985). Coevolution is dynamic, proceeding from coexistence to absolute interdependence, and must be viewed over millennia rather than decades (Greenwood, 1981). Commonly studied examples, such as pollination, seed dispersal, herbivory, and mimicry, do not provide examples of intimacy or integration. Such associations are either too recent or, more often, represent coevolved equilibria whose functionality depends on the biological and physical autonomy of the symbionts.

The more intimate and stable a symbiosis, the greater is the probability of it being the outcome of coevolution. Both strongly positive (mutualistic) and strongly negative (antagonistic) interactions "provide the selective thrust for coevolutionary development (Southwood, 1985), and "coevolution almost unavoidably must occur once an intricate relationship has been established" (Parlevliet, 1986). As the basis of mutualism appears to be "reciprocal exploitation rather than the anthropomorphic concept of symbionts 'acting for each other's benefit' " (Barrett, 1983), mutualistic symbioses probably evolved from initially antagonistic relationships and may well be the outcome of coevolution.

With an appreciation of the coupling of coevolution with symbiosis, especially when mutualistic, persistent, intimate, and stable, increasing attention is being given to microorganisms (i.e., monera, protoctista and fungi) that interact multifariously among themselves and mediate a plethora of plant–animal interactions (Jones, 1984). "When we look more closely at plant–animal trophic links we see that this Kingdom to Kingdom link . . . is an over-simplification. There is a third, a vital member of the relationship, a microorganism—a member of one of the three other Kingdoms . . . they have in the course of evolution become the 'biochemical brokers' in the plant/animal interaction" (Southwood, 1985).

To mycologists the concept of coevolution is not new but taken for granted. Most fungi are obligate and often mutualistic symbionts. Since Dietel (1904) reported reciprocal interactions between rusts and their hosts, and despite the numerous studies of Savile (see 1976, 1979), the subject attracted little attention outside mycology (Wahl *et al.*, 1984). Admittedly, some uredinologists interpreted coevolution as the evolution of a parasite in conjunction with the evolving host—"Host plants evolve and the pathogens coevolve with them" (Parlevliet, 1986)—but this is not what coevolution *sensu stricto* now signifies. However, plant pathologists also provided a genetic model of symbiosis that helped to formulate the modern restricted definition of coevolution: the "gene-for-gene" relationship as in the flax–rust symbiosis (Flor, 1955; Christ *et al.*, 1987). The gene-for-gene system is one of the genetic mechanisms of the sociologists' "arms race": as new mutations arise or new genes are introduced, as in the breeding of crop plants for resistance, a prey or host species evolves or acquires new defences while the predator or parasite evolves new ways of overcoming them. Parasitologists are well aware of the evolutionary significance of reciprocal adjustments in host–parasite associations, including the genetic polymorphism they appear to generate (Burdon, 1985). They pioneered principles and methodologies for applying coevolutionary criteria to systematics, phylogenetics and biogeography (Brooks, 1979; Price, 1980; Price *et al.*, 1986a; Stone and Hawksworth, 1986) to which mycologists are also established, albeit less explicit, subscribers (Savile, 1979; Hedberg, 1979).

Perhaps because fungi maintain intimate links with their living substrata, mycologists tend to take coevolution for granted. The result is that mycology now lags behind other disciplines in this field of study. To claim that some kind of biological reciprocity in fungal symbioses must invariably occur is one thing; to demonstrate coevolution in action is quite another. "The inference that coevolution in plant–fungus symbioses has occurred is a plausible and probably the most parsimonious explanation of their existence" (Barrett, 1983). Barrett also stresses how limited are the empirical data and how difficult they are to verify even in such well-publicized mutualistic symbioses as lichens and mycorrhizas. However, in this volume we have deliberately ignored Barrett's advice to shun "adaptive storytelling". If it had been taken, this book might not have materialized. By recording the array of diverse symbioses with fungi, we aim to identify associations deserving more vigorous appraisal by students of coevolution. It is our contention that attention should be focused on the scale and variety of coevolutionary events involving fungi.

Fungi can be expected to have coevolved, and indeed to be coevolving, with representatives of all forms of life from viruses to man. In interacting with other organisms, fungi invade cells or tissues in search of nutrients, shelter, or transport, or mediate the interface between these organisms and their physical and biological environment in exchange for nutrients, shelter, or transport. The true fungi (Eumycota) lack phagocytosis and "naked" motile stages and are remarkably free of endosymbionts. The exceptions are the viruses, the unidentified bacteria-like inhabitants of hyphae of some vesicular-arbuscular mycorrhizal fungi, the enigmatic "concentric bodies" in some lichenized and free-living ascomycetes, and mycoparasitic fungi.

COEVOLUTION OF FUNGI WITH VIRUSES, MONERANS, PROTOCTISTS AND FUNGI

With the exception of lichens (Hawksworth, this volume, Chapter 6), the treatment of these groups is peripheral to the theme of this book and is here mentioned only briefly for completeness.

Mycoviruses

Lemke (1979) regarded the symptomless association of double-stranded RNA viruses with fungi as representing a coevolved mutualism in which the host benefits from survival and the virus from persistence. The fungus exercises polygenic control over expression of viral sequences such as those encoding for determinants of the "killer" phenotype. Conversely, mycoviruses may contribute part of the genetic information for the synthesis of certain secondary metabolites by host fungi or contribute to modification of virulence (Buck, 1987).

Bacteria

Fungi and bacteria do not provide "textbook" examples of intimacy or integration, but symbioses between them appear to be common, are potentially ancient and, consequently, could be significant from a coevolutionary standpoint. Coevolution, however, remains largely to be demonstrated. Fungi and bacteria compete directly for the limited range of assimilable nutrients and have evolved a formidable arsenal of chemical weaponry for this purpose. "Chemical warfare" need not curtail reciprocal change, but mutual evolution of virulence and resistance is the likely outcome of prolonged interaction.

factor responsible for the diversification of the present-day parasites; the others are (a) topographical specialization in relation to host tissues; (b) biological specialization to a range of hosts via evolution of the parasite in conjunction with the evolving host; (c) sporadic successful "jumps" to a more distantly related host (Parlevliet, 1979, 1986; Hijwegen, this volume, Chapter 3). However, establishment of greater physical proximity between the interacting cells, increased host specialization through reciprocal evolution of virulence and resistance to toxins, or fine-tuning of parasite to host after opportunistic colonization, may all lead to coevolution even if they do not involve that initially. It is not surprising, therefore, that hypotheses explaining the logistics of fungus–plant interactions range from postulation of indirect responses of plants, to mechanical damage caused by fungi, to direct altering by fungi of the hosts' gene expression (Heath, 1986). Undoubtedly "there is dynamic genetic interactions between host and fungal parasite" (Barrett, 1983) in the form of gene-for-gene relationships. Nevertheless, this model of reciprocal microevolutionary change rests on examples from grossly disturbed agrosystems and may not be representative of natural ecosystems. Indeed, the gene-for-gene model does not apply to all agriculturally manipulated systems. Browning (1981) stresses that such gene pairs are natural, not a man-made artifact, but admits that the epidemiological effect of selective filtering of incoming fungal inoculum may be the result of agriculture. Plant pathology and parasitology consequently provide direct evidence, even if distorted by man, of the reciprocal microevolutionary change that underlies coevolution.

No comparable microevolutionary data exist for another conspicuous and consequential plant–fungus association, mycorrhizas, mycothalli, and mycorrhizomes (Boullard, this volume, Chapter 5); nor do they exist for those involving fungi and non-economic plants. Mycotrophic symbioses alter the morphology and physiology of the plant. Increased assimilation rate of nutrients by the plant may merely compensate for the loss of energy to the fungus, but benefits accrue from increased resistance to or tolerance of biotic and abiotic stresses. The fitness of the associating organisms becomes greater when together than when apart; to Law and Lewis (1983) this signals mutualism. The physiological interplay in mycorrhizas has proved difficult to demonstrate. Genetic interactions have not been investigated, though hints of microevolutionary adjustments may be found (Gianinazzi-Pearson and Gianinazzi, 1986). Trappe and Molina (1986) note, following Person (1968), that "selection cannot favour a gene whose only effect is to benefit another organism", and echo Barrett (1986) in granting that stepwise mutations may lead to mutually beneficial reciprocal exploitation; such "secondary" benefit characterizes mycotrophic symbioses. However, almost nothing is known as to whether genetic interaction or integration is involved in the biological adjustment of fungi and plants or in their joint response to the external environment, both physical and biotic. Extracellular growth substances produced by ectomycorrhizal fungi induce mycorrhizal morphology and may account for the initiation of mycorrhizal symbioses, and for physiological changes in the host in relation to nutrient flow to the fungus. However, comparable morphological change in the root can be induced by cell-free extracts of appropriate fungi, and even by autonomous saprobic microorganisms in the rhizosphere (Tinker, 1984) or inorganic oxidizing agents such as hydrogen peroxide. Faye *et al.* (1980) concluded that the formation of mycorrhizogenic short roots is under the genetic control of the plant. Although the presence or absence of symbiotic fungi alters the susceptibility of normally ectomycorrhizal plants to infection by biotrophic fungi, genetically defined resistance to disease remains unchanged (Meyer and Dehne, 1986). When a root is protected by an ectomycorrhizal fungus, resistance to infection may be largely due to the production of antibiotics by the fungus rather than the induction of resistance mechanisms in the host.

It has been suggested that, as in lichens (Hawksworth, this volume, Chapter 6), ectomycorrhizal symbiosis is a controlled parasitism of a plant by the fungus or vice versa (Malloch, 1988). The boundary between antagonism and mutualism (the likely product of coevolution) is precarious and unstable, but we feel it is most probable that the ectomycorrhizal "stand off" is a coevolved mutualism, as is the case in lichens (see Chapter 6).

COEVOLUTION OF FUNGI WITH ANIMALS

Interactions between fungi and animals are as prevalent, ancient, and evolutionarily consequential in, for example, promoting diversity, as those between fungi and plants, and they have been equally poorly studied from a coevolutionary perspective. Antagonistic symbioses range from destructive invasion by fungi (exploited and with latent potential in the biocontrol of nematode and insect pests; Evans, this volume, Chapter 7), to probably commensalistic associations such as between insects and the Laboulbeniales (Evans, Chapter 7). The heterotrophic hosts of fungi range from amoebae to man. Many interactions lethal to invertebrates secure a store of nutrients through rapid immobilization and killing, sometimes accompanied by preservation of the resource by toxins, mechanical means, or both (Cooke, 1977; Charnley, 1984; Ahmadjian and Paracer, 1986). Others, depending on liaison with a living and functioning animal host, appear to be engaged in an "arms race" conflict, circumventing defences from simple mechanical barriers to complex immunological responses fine-tuned by coevolution.

We are impressed with the bizarre examples from the parasitological litera-

lium (Martin, 1984). Siricid wood wasps similarly obtain enzymes for wood degradation from "cultivated" white rot fungi *Stereum* and *Amylostereum*. Conidia are introduced with eggs through a specially modified ovipositor. The fungi grow in galleries and, mixed with wood, serve as larval food (Martin, 1984; Gilbertson, 1984). These authors consider the "borrowing" of fungal enzymes by arthropods and other invertebrates probably to be widespread and coevolved. In the case of siricid wasps and similar examples, this "coevolutionary adjustment" is particularly conspicuous in the adaptations of the insect; unlike the fungus, the wasps cannot complete their life cycle independently. However, transport and entry considerations must be beneficial to the fungus, and the specialized conidial oidia may result from the coevolution.

The "mycetome" may also be a product of coevolution in insects. Many insects contain symbiotic organisms within specialized cells ("mycetocytes") that may be scattered or aggregated into a coherent structure, the mycetome. Contrary to the implication in the name, the chief inhabitants of mycetomes are prokaryotes not filamentous fungi, although yeasts sometimes occur. Insects with such symbionts can live on nutritionally poor or unbalanced diets such as keratin, plant sap, or blood, obtaining sterols, nitrogen, and vitamins from the symbionts. The host appears to have some control over the metabolism of the symbiont. Mycetocyte symbiosis may have evolved from gut symbioses (Smith and Douglas, 1987). In anobiid beetles the endosymbionts are yeasts, occurring intra- and extra-cellularly in mycetomes attached to the larval hindgut (Crowson, 1984); gut outpocketing has been claimed to have led to the evolution of mycetomes.

COEVOLUTION OF FUNGI WITH PLANTS AND ANIMALS

In the preceding section reference was made to fungi that transform nonliving organic substrata, such as the components of wood not directly assimilable, or toxins, into products acceptable or indispensable to the animal with which they are associated. It can be assumed, therefore, that fungi mediate to an even greater extent in interactions between animals and their food supply, i.e. responding plant hosts. The assumption is necessary in many of the following examples; the "more complex tripartite interactions involving higher plants, microbial pathogens and their invertebrate vectors, and those between vertebrates, pathogens and vectors are only elementary examples of the potential multiplicity of relationships within natural systems" (Anderson *et al.*, 1984).

In multilevel symbioses, two or more organisms combine resources to exploit more effectively another organism or symbiotic system, or all interact to the mutual benefit of enhanced fitness in a novel or marginal environment. Since fortuitous symbioses provide an opportunity for coevolution, and the established ones are fine-tuned by it, the examples of symbiosis below are conventionally classified as antagonistic or mutualistic, but the boundaries are by no means clear. As Smith and Douglas (1987) stress, it is difficult to quantify mutualism even in stable one-to-one associations. A casual, qualitative appraisal of the nature of interaction in more complex tripartite or multipartite associations can lead to erroneous conclusions. There is a danger of anthropocentric bias. For example, a *Vaccinium* berry mummified by *Monilinia* (Batra, 1987) that shrivels and falls to the ground is obviously the outcome of an antagonistic symbiosis if evaluated in terms of the loss of economically important fruit and measured in dollars. However, what is significant to this plant is not who destroys the fruit and how, but the fate of the seed; a seed on the ground is better than one in a blueberry pie.

The "real world" picture of symbiosis is enormously complex, and we cannot begin to evaluate coevolution with fungi in the light of present knowledge, except as an anecdotal narrative. However, we anticipate that when the direction and extent of coevolution in multilevel symbioses is more fully elucidated, some apparently "far-out" interpretations will turn out to be conservatively pedestrian.

In the following examples of tripartite symbioses with fungi, most plants and animals are not individuals but symbiotic systems: plants with mycorrhizal fungi, animals with gut microorganisms that include fungi. Further, each biont interacts with its own coterie of fungal and animal parasites, pathogens, dispersers, etc.

Mycorrhizal fungi

Relationships of plants and mycorrhizal fungi were noted above (p. 8). "Real world plant nutrition" (Tinker, 1984) always occurs in a complex system in which microbial interactions are "probably the most important, yet the least understood, phenomena in rhizosphere biology" (Bowen, 1980). Mycorrhizal fungi are only the first, albeit the most important, level of symbionts of the interface not only between a plant and its physical environment as the source of essential minerals, but also between it and the biological environment, including other fungi, bacteria, protoctists, and animals. Some symbioses are antagonistic, and others mutualistic (synergistic), presumably involving coevolution such as might be suspected to occur between plants, VAM fungi, and nitrogen-fixing prokaryotes (Daft *et al*., 1985). Interaction between mycorrhizal and other fungi, both as rhizosphere competitors and root and shoot parasites, is an increasingly popular field of study first pioneered by Soviet scientists (Bagyaraj, 1984; Lindeberg, 1986). In particular, "the possible role of soil microfauna is an important omission from most studies on microorganism–plant interactions" (Bowen, 1980), though recent interest in the interplay between VA mycorrhizal fungi and nematodes (Bagyaraj, 1984) is redressing the imbalance. It is clear that many heterotrophic protoctists and animals (in addition to nematodes), eat mycorrhizal fungi, while others may be "eaten" by mycorrhizal fungi or their close relatives (Sagara, 1981). Case studies are still few and far between. Even more neglected are interactions between mycorrhizal fungi and vertebrates, especially in relation to zootic dispersal of seeds (Pirozynski and Malloch, this volume, Chapter 10).

Gut microorganisms

Gut microorganisms can transform, supplement, or detoxify animal food (Ahmadjian and Paracer, 1986; Smith and Douglas, 1987). Gut symbionts appear to adapt to novel substances by virtue of a flexibility in community structure as well as the adaptability of metabolic pathways in specific populations (Levin, 1976). Interactions resulting in the detoxification of plant metabolites consumed by herbivores may involve one-to-one "arms race" coevolution between gut symbionts and the plant. Conversely, coevolution of a consumer with its gut microbiota is more likely to be "coarse" or "diffuse" (Herrera, 1985; Janzen, 1985), in that there is pronounced asymmetry in the turnover (evolutionary rates) of microbial species, variable composition of microbial communities, and a low level of dependence on the host animal. Microbial gut symbionts were probably originally associated with plants or plant organs, acquired during feeding, and subsequently deployed to detoxify defence metabolites of their former hosts (Rosenthal, 1983). In view of their versatility and effectiveness in detoxification of a range of plant products, including phytoalexins, alkaloids, and cyanides, fungi could be potentially desirable as symbionts and may play a greater role in gut symbioses than is currently recognized. This is the case in the conversion of polysaccharides by ruminants (Bouchop, 1981; Williams and Orpin, 1987).

Superimposed on mycorrhizal and gut symbioses are other plant–fungus–animal associations. They are also seldom halves of independently evolving or coevolving pairs: the aim of heterotrophs is to obtain energy from some plant product that is sequestered by one (in our examples usually the fungus) and passed on or shared with the other in exchange for shelter, "comfort", or dispersal. As heterotrophs, both fungi and animals ultimately depend on photosynthetically fixed carbon. Competition aims at tapping this limiting resource close to the source (i.e., living plants). Consequently, in tripartite interactions between plants, fungi, and animals, the plants often appear not to benefit or to be adversely affected; this impression is reinforced by the situation in man-made or manipulated agroecosystems. In nature, coevolution has frequently mellowed combat into truce, and symbioses that at first sight appear antagonistic may, on closer inspection, reveal cryptic benefits to the plant.

Septobasidium

In this example of plant–insect–fungus symbiosis (Evans, Chapter 7), the disadvantaged biont appears to be the plant, which suffers mechanical injury by the insect's mouth parts, and loss of sap. In this instance, unlike the following, it is the insect that interacts more directly with the plant and has evolved morphological and physiological adaptations for sedentary habit, a diet of sap, topographical requirements, and host specificity; some of these adaptations could have been fine-tuned by coevolution. This more obvious first impression of the first level interaction may, however, be deceptive if the insect injects fungal metabolites, as has been indicated by Cooke (1977). The fungus primarily parasitizes individual insects, immobilizing, dwarfing them, and making them sterile albeit longer-lived. However, because the insects are gregarious and the fungus shelters a population of potential hosts, not all individuals are attacked. From the evolutionary perspective, only the uninfected individuals benefit, and they in turn benefit the fungus by providing the sole means of spore dispersal. From the coevolutionary perspective, the relationships are more difficult to identify. It would seem that it is the "sacrificed" individual that is the site of insect–fungus coevolution and it would be pertinent to identify what controls the ratio of infected to uninfected individuals within a colony, or the pattern of fungal growth that results in the formation of a protective stroma. A parallel case wherein the chief beneficiary of mutualistic association of a plant and its fungal endophyte is not the infected individual but the uninfected one, which benefits from reduced population of herbivores, was discussed by Carroll (1986; see also Evans, this volume, Chapter 7). This explanation impinges on the controversial concepts of kin selection and altruism, discussions of which are premature with regard to fungi.

Sooty moulds

In some other stable associations in which the insect is at the trophic interface between plant and fungus, the fungus does not invade the haemocoel but rather derives energy from underutilized photosynthates excreted by the insect. Sooty moulds growing on sugary honeydew exudates of sap-sucking insects are an example. It would seem that the interaction between plant and fungus is sufficiently indirect to preclude coevolution. However, Patelle (1982) suggested that the plant may not be the exploited member of the trio because sap-sucking insects could "act as a mechanism for plants to provide carbon and energy source to a microbial community that can benefit the plant", both in the rhizosphere and the phyllosphere. Furthermore, far from considering sooty moulds as "cheaters" of the system, he regarded them as a component of a coevolved, mutualistic package providing the plant with a means of controlling photosynthesis and transpiration rates and, more directly (and credibly), offering protection against herbivores and parasites.

Another example of a fungus benefiting from herbivory is seen in Attini ants, though in this case it is more difficult to imagine a coevolved connection between the plant and the fungus, even if mycorrhizal fungi paved the way to ant–fungus mutualism (Garling, 1979).

Animal vectors of plant parasitic fungi

It is beyond the scope of this review to include the voluminous literature pertinent to the transmission of plant parasitic fungi by animal vectors. From the symbiotic point of view, however, the plant is usually disadvantaged, especially if a cross section of the literature, which is heavily biased towards agroecosystems, is examined. It is the fungus and its animal vector that are more likely to be mutualistically linked, and consequently coevolved. The association of *Epichloë typhina*, the agent of "choke" disease of grasses, and the fly *Phorbia phrenione* (Kohlmeyer and Kohlmeyer, 1974) is an example (p. 20). The damaging, antagonistic behaviour of a pathogenic fungus need not preclude coevolution with the plant host, even when the weapons are toxins rather than supremacy in gene-for-gene strategy. Even prolonged, intimate associations are not static (Van Valen, 1973); their deceptive "stability" is balanced by coevolutionary adjustment. Furthermore, behind the facade of conspicuous antagonism may lie a benefit to the plant: many grasses infected by endophytic clavicipitaceous relatives of *Epichloë typhina* produce greater quantities of more viable seeds than the uninfected plants (Clay, 1987). A less enigmatic benefit to host plant of infection was claimed by Jennersten (1983). Pollinating butterflies acting as specific vectors of anther smut (*Ustilago violacea*) of some Caryophyllaceae are not only unwitting benefactors of the fungus whose spores they carry to the appropriate organ of the correct plant, but also benefit the plant indirectly via the fungus. In sterilizing the anthers of hermaphrodite flowers, the fungus promotes cross-pollination, and so inherent long-term advantages to the host population.

A more elaborate case was investigated by Batra and Batra (1985) who reported that the discomycete *Monilinia vaccinii-corymbosi*, which attacks the leaves of blueberries and huckleberries, competes with the flowers for pollinating insects. By inducing changes in both the visible-light and ultraviolet colour patterns of leaves, secretion of sugars, and the emission of fragrance, the fungus entices opportunistic pollinators to forage on conidia-covered leaves and thus ensures transport to flowers that are the site of mycelial overwintering.

A similar deception is practised by rust fungi in that, according to Leppik (1974, 1975), some pycnia attempt to mimic flowers by (a) offering nitrogenous food corresponding in composition to pollen, (b) slime, sugar and water, corresponding to nectar, (c) white, yellow or orange pigmentation, and (d) odour that attracts insect pollinators. Although "insects take an active part in the life and evolution of rust fungi" (Leppik, 1975), if coevolution is involved in this and the preceding example it is more likely to operate in microevolutionary adjustment of the plant and its fungal parasite. However, if mimicry in the form of one system parasitizing a previously evolved or coevolved system engaged in pollination mutualism proves to be the case, all three components may be the result of coevolution (Gilbert, 1983). Although there are no compelling reasons to suggest that the product of plant–fungus coevolution mimics that of plant–insect coevolution, or that the plant benefits from infection and cooperates to secure it, this idea must not be summarily dismissed. One possible advantage to the host of maintaining a population of parasites is their preventing immigration of more susceptible competitors (Barrett, 1986). Heteroecism in rust fungi was claimed (Rice and Westoby, 1982) to have evolved as an agency of interference competition between the two hosts; however, Savile (1976) regarded it as having originated under severe environmental stress caused by aridity.

Ambrosia galls

Some plant galls, especially those caused by cecidomyiid midges, have a third crucial component, which is a fungus. The association of the insect with the fungus appears to be mutualistic: the larvae eat the mycelium, and the adults disperse spores. The insect obtains a steady supply of highly compatible, "farmed" food, and the fungus is dispersed into a highly specialized, protected niche, and perhaps also extends its normally saprotrophic habit into biotrophy. That coevolution may have taken place is indicated by the physical traits (mycangia) and behavioural traits of the insect for collecting and carrying specific fungal spores used in inoculating specific plant tissue. Furthermore, the insect appears to control physiological processes and the morphogenetic expression of the fungus, for example the timing of spore ger-

man *et al.*, 1986). Wicklow and Rebar (1988) showed that banner-tailed kangaroo rats carry in cheek pouches a complement of spores that varies in composition according to seasonal temperature regimes, and with which freshly harvested seeds are inadvertently "inoculated". These animals preferentially feed on slightly mouldy seeds (Reichman *et al.*, 1985). The advantages to the animal are not clear, but the rodents may benefit from the acquisition of antimicrobial antibiotics that the grain moulds produce. Alternatively, being larger and being mammals, these animals may be able to tolerate toxin levels that deter arthropod competitors, such as bruchid beetles (D. T. Wicklow, personal communication). The advantages to the fungus in securing food, shelter, and transport are obvious. With respect to the plant, seeds, even when mouldy, remain viable, and seed-caching rodents are a major agency of seedling recruitment (Pirozynski and Malloch, this volume, Chapter 10). Two additional observations may be pertinent: (a) chemically well-protected seeds generally do not become mouldy in rodent caches, and (b) a significant proportion of grain fungi are proteolytic (D. T. Wicklow and D. W. Malloch, personal communication). Perhaps the biological protection of seeds by moulds is a coevolved ecological strategy for seeds that, like those of grasses and many legumes, are not protected chemically but pay for protection in proteins of seed test or aleurone.

Endophytes and herbivores

Endophytic fungi have been found in all phanerogams so far investigated, to be an integral aspect of these plants' ecological adaptations (Petrini, 1984, 1985; Carroll, 1986). That plants asymptomatically infected by systemic fungi are "fitter" than uninfected plants has been best demonstrated for grasses, sedges, and their clavicipitaceous endophytes (Clay, this volume, Chapter 4). One of the benefits to the plant lies in the deployment of fungal metabolites as deterrents to insect and mammalian herbivores (Siegel *et al.*, 1987). In turn, the fungus obtains a secure supply of food, shelter, and ultimately incorporation in the hosts' seed (Cole and White, 1985; Philipson and Christey, 1985). However, before so convenient an arrangement can be implemented (and in woody plants it has not (Carroll, 1986)) the fungus must rely on more demanding and fickle animal vectors. In the case of *Epichloë typhina*, the fly *Phorbia phrenione* is attracted to the conidiating stroma and feeds on conidiophores and conidia, which action stimulates oviposition on the fungal stroma. The conidia pass through the digestive tract in a viable condition to infect new individuals of a potential host (Kohlmeyer and Kohlmeyer, 1974). Kohlmeyer and Kohlmeyer considered the insect to be parasitic on *Epichloë*, but the interaction is more likely to be mutualistic and possibly coevolved. A more bizzare ramification of presumably the same evolutionary story concerns fungi: the Clavicipitaceae are noted entomopathogens (Evans, this volume, Chapter 7).

Conversely, and in apparent conflict of interests, there are reports of purportedly mutualistic, coevolved associations between grasses and grazers (Belsky, 1986; Verkaar, 1986) based on observations of grazed plants gaining in biomass and "fitness".

With respect to the chemical defences of plants "one herbivore's poison is another herbivore's food" (Janzen, 1981), i.e. there usually is an animal that has acquired the ability to detoxify and, therefore, more or less monopolize a particular source of food (Rosenthal, 1983) or even sequester the active ingredients for its own protection (Harborne, 1982, p. 87). Fungi have the capacity to produce particularly toxic compounds (Moss, 1984), but are not themselves immune to exploitation. For example, ergot sclerotia are eaten by a phalacrid beetle (Wicklow, this volume, Chapter 8), and mycophagous drosophilids both tolerate α-amanitin (Jaenike *et al*., 1983) and benefit from it in controlling abdominal parasitic nematodes (J. Jaenike, personal communication).

Nutritional specialization of an animal to a specific plant-borne fungal product is not likely to account for the purported coevolution of grasses with relatively generalistic grazers, but it may play a role in specialized one-to-one examples of zoochory. While there is no problem in interpreting the *Septobasidium* trio (p. 15), where only some insects of a colony are parasitized, as mutualistic, we fail to see why, for example, a covered smut that regularly affects only some seeds in a fruit, is anything but straightforward parasitism. However, it is theoretically possible that a plant so infected could gain from the selective harvesting of fruits and dispersal of seeds to a safe site: the animal from being directed to food that may also be protected against opportunistic generalist competitors; and the fungus from being defaecated or cached with seeds—this could be the next best alternative to being seed-borne and does not carry the penalty of suppressed sexuality (Clay, Chapter 4; Pirozynski, Chapter 11, both this volume).

The Myrmecophiles

In some ant-associated epiphytes, particularly *Myrmecodia* and related Rubiacae, ants live in tuberous chambered modifications of the hypocotyl, keeping brood in chambers having smooth walls and depositing most of their faeces and other organic debris in chambers with warted, absorptive walls. An *Arthrocladium*-like hyphomycete is typically associated with warted chambers and reputedly renders organic debris assimilable by the plant (Huxley, 1978, 1982).

In this tripartite symbiosis, all participants visibly benefit from the association: (a) the plant is "planted" and supplied with essential nutrients in a situation in which it escapes the rhizosphere and ground cover competitors, and may also be actively protected by the ants against herbivores; (b) the fungus is supplied with shelter, which may also provide a physical environment suitable for growth, uniformly processed nutrients and transport; (c) the fungus appears to be dependent on ants for dispersal and "planting" in appropriate chambers of appropriate hosts; and (d) similarly, the ants obtain shelter from the plant and nutrient from the fungus on which they feed.

Superficially, this neat cooperative packaged in a bizarre morphological chimera is the epitome of coevolution. However, if the plant is the product of coevolution, the corresponding symbiont or symbionts may be extinct or no longer recognizable as organisms, and it or they may have been other than fungi or ants. The chambered "tuber" forms when the plant is grown from seed irrespective of whether fungi or ants are present. To use plant-gall terminology, which here seems appropriate, the ants and the fungi may be merely *successori*. It does not mean that coevolution between the plant and its current occupants is not possible or even probable. However, not all, and not even the most conspicuous traits that make the system work to mutual benefit of plants, fungi, and ants must be of their making.

It seemed appropriate to close this introductory chapter with a potential case of sequential adaptations, for it may occasion reflection on the examples already cited and interpretations of those that follow in later contributions from the perspective of time.

THE SIGNIFICANCE OF COEVOLUTION INVOLVING FUNGI

In this contribution, we have endeavoured to introduce the concept of coevolution as it pertains to the fungi, and more particularly to provide an overview of the situations in which there are indications that it has been of significance in shaping the course of evolution in both fungi and other organisms with which they interact. In the majority of cases, the evidence remains largely circumstantial, but even if only a proportion are vindicated by future more critical work, the phenomenon is evidently so widespread and fundamental that it can no longer be ignored in more general considerations of evolution of both fungi and the groups of plant and animals involved.

Although mycologists, for example, Savile (1954), recorded coevolutionary phenomena long before the term was coined by Ehrlich and Raven (1964), the phenomenon is generally implied rather than explicitly discussed in mycological texts (e.g., Batra, 1979; Wicklow and Carroll, 1981; Anderson *et al.*, 1984; Wheeler and Blackwell, 1984). Even literature on mycorrhizas, mycotrophic symbioses, while recognizing the mutualistic aspects, stops short of considering reciprocal evolutionary implications (Harley and Smith, 1984). Coevolution is scarcely mentioned in the survey of evolutionary biology of the fungi compiled by Rayner *et al.* (1987). In this context, contributions involving fungi included in Hedberg (1979) provide a stimulating exception.

The result is that, in the field of coevolution, the role of fungi is vastly underestimated and understudied. Together with the succeeding chapters in this volume, we have attempted to focus attention on the fungi by discussing a wide range of interactions with plants and animals in a mycological context. In the years ahead, we trust, many of the cases identified here will be subjected to penetrating scrutiny to identify more clearly reciprocal changes at the genotypic level in the way only yet achieved for certain pathogens of crop plants (Barrett, 1983, 1986; Parlevliet, 1986; Hijwegen, this volume, Chapter 3).

The comparison of cladograms constructed independently for fungi and the interactive organisms may be particularly instructive in future analyses. This approach was adopted by Humphries *et al.* (1986) in relation to *Cyttaria* species and their *Nothofagus* hosts, although Fahrenholz's rule that the phylogeny of parasites mirrors that of their hosts has not been substantiated in several studies (see Stone and Hawksworth, 1986). However, it must be borne in mind that the real world of coevolution is not pairwise between individuals but rather between symbiotic systems.

Mycologists also need to contribute to more general debates on coevolution. Fungi receive scant attention in recent theoretical examinations of this subject (e.g., Futuyama and Slatkin, 1983; Thompson, 1982, 1986); this situation will only be redressed by mycologists taking an interest in and attempting to assess the relevance of concepts such as those of "mixed-process coevolution" and "interaction norms" (Thompson, 1986).

There are exciting new fields open for original research in this hitherto neglected area of evolutionary biology. If we have stimulated enquiry and debate into these, our objective will have been realized.

ACKNOWLEDGEMENTS

We are indebted to Drs L. R. and S. W. T. Batra for supplying an unpublished synopsis of their fascinating work on floral mimicry in blueberries. In addition, we benefited from Dr D. B. O. Savile's constructive comments on an early draft of this contribution.

REFERENCES

Ahmadjian, V. and Paracer, S. (1986). "Symbiosis". University Press of New England, Hanover.

Anderson, J. M. and Ineson, P. (1984). Interactions between microorganisms and soil invertebrates in nutrient flux pathways of forest ecosystems. In "Invertebrate–Microbial Interactions" (J. M. Anderson, A. D. M. Rayner, and D. W. H. Walton, eds), pp. 59–88. Cambridge University Press, Cambridge.

Anderson, J. M., Rayner, A. D. M., and Walton, D. W. H. (eds) (1984). "Invertebrate–Microbial Interactions". Cambridge University Press, Cambridge.

Ashe, J. S. (1984). Major features of the evolution of relationships between Gyrophaenine staphylinid beetles (Coleoptera Staphylinidae: Aleocharinae) and fresh mushrooms. In "Fungus–Insect Relationships" (Q. Wheeler and M. Blackwell, eds), pp. 227–255. Columbia University Press, New York.

Bagyaraj, J. (1984). Biological interactions with VA mycorrhizal fungi. In "VA Mycorrhiza" (C. L. Powell and J. Bagyaraj, eds), pp. 131–153. CRC Press, Boca Raton, Fla.

Barnett, H. L. (1968). The effects of light, pyridoxine and biotin on the development of the mycoparasite, *Gonatobotryum fuscum*. *Mycologia* **60,** 244–251.

Barnett, H. L. and Binder, F. L. (1973). The fungal host–parasite relationship. *A. Rev. Phytopath.* **11,** 273–292.

Barrett, J. A. (1983). Plant–fungus symbioses. In "Coevolution" (D. J. Futuyma and M. Slatkin, eds), pp. 137–160. Sinauer Associates, Sunderland, Mass.

Barrett, J. A. (1986). Host–parasite interactions and systematics. In "Coevolution and Systematics" (A. R. Stone and D. L. Hawskworth, eds), pp. 1–17. Clarendon Press, Oxford.

de Bary, A. (1879). "Die Erscheinung der Symbiose". Trübner, Strasbourg.

Batra, L. R. (ed.) (1979). "Insect–Fungus Symbiosis". Wiley, New York.

Batra, L. R. and Batra, S. W. T. (1985). Floral mimicry induced by mummy-berry fungus exploits host's pollinators as vectors. *Science, N.Y.* **228,** 1011–1012.

Batra, S. W. T. (1987). Deceit and corruption in the blueberry patch. *Nat. Hist.* **96,** 57–59.

Belsky, A. J. (1986). Does herbivory benefit plants? A review of the evidence. *Am. Nat.* **127**, 870–892.

Bouchop, T. (1981). Anaerobic fungi in rumen fibre digestion. *Agric. Envir.* **6,** 339–348.

Bowen, G. D. (1980). Misconceptions, concepts and approaches in rhizosphere biology. In "Contemporary Microbial Ecology" (D. C. Ellwood, J. N. Hedger, M. T. Latham, J. M. Lynch, and J. H. Slater, eds), pp. 283–303. Academic Press, London.

Brazner, J., Aberdeen, V. and Starmer, W. T. (1984). Host–plant shifts and adult survival in the cactus breeding *Drosophila mojavensis*. *Ecol. Entomol.* **9,** 375–381.

Brodo, I. M. (1976). Lichenes Canadenses Exsiccati: Fascicle II. *Bryologist* **79,** 396–399.

Brooks, D. R. (1979). Testing the context and extent of host–parasite coevolution. *Syst. Zool.* **28,** 299–307.

Browning, J. A. (1981). The agro-ecosystem—natural ecosystem dichotomy and its impact on phytopathological concepts. In "Pests, Pathogens and Vegetation" (J. M. Thresh, ed.), pp. 159–172. Pitman, Boston.

Buck, K. W. (1987). Viruses of plant pathogenic fungi. In "Genetics and Plant Pathogenesis" (P. R. Day and G. J. Jellis, eds, pp. 111–126. Blackwell Scientific Publications, Oxford.

Burdon, J. J. (1985). Pathogens and genetic structure of plant populations. In "Studies on Plant Demography" (J. White, ed.), pp. 313–325. Academic Press, London.

Carroll, G. C. (1986). The biology of endophytism in plants with particular reference to woody perennials. In "Microbiology of the Phyllosphere" (N. J. Fokkema and J. van der Heuvel, eds), pp. 205–222. Cambridge University Press, London.

Charnley, A. K. (1984). Physiological aspects of destructive pathogenesis in insects by fungi: a speculative review. In "Invertebrate–Microbial Interactions" (J. M. Anderson, A. D. M. Rayner, and D. W. H. Walton, eds), pp. 229–270. Cambridge University Press, Cambridge.

Christ, B. J., Person, C. O., and Pope, D. D. (1987). The genetic determination of variation in pathogeneticity. In "Populations of Plant Pathogens" (M. S. Wolfe and C. E. Caten, eds), pp. 7–19. Blackwell Scientific Publications, Oxford.

Clay, K. (1987). Effects of fungal endophytes on the seed and seedling biology of *Lolium perenne* and *Festuca arundinacea*. *Oecologia* **73,** 358–362.

Cole, G. T. and White, J. F. (1985). Fungal endophytes of forage grasses. *Proc. Indian Acad. Sci. (Plant Sci.)* **94,** 129–136.

Cooke, R. (1977). "The Biology of Symbiotic Fungi". Wiley, London.

Crowson, R. A. (1984). The association of Coleoptera with ascomycetes. In "Fungus–Insect Relationships" (Q. Wheeler and M. Blackwell, eds), pp. 256–285. Columbia University Press, New York.

Daft, M. J., Clelland, D. M., and Gardner, I. C. (1985). Symbiosis with endomycorrhizas and nitrogen-fixing organisms. *Proc. R. Soc. Edinb.* **85B,** 283–298.

Dietel, P. (1904). Betrachtungen über die Verteilung der Uredineen auf ihren Nährpflanzen. *Zentbl. Bakt. ParasitKde*, Abt. II **12,** 218–234.

Dowding, P. (1984). The evolution of insect–fungus relationships in primary invasion of forest timber. In "Invertebrate–Microbial Interactions" (J. M. Anderson, A. D. M. Rayner, and D. W. H. Walton, eds), pp. 133–153. Cambridge University Press, Cambridge.

Ehrlich, P. R. and Raven, P. H. (1964). Butterflies and plants: a study in coevolution. *Evolution* **18,** 586–608.

Eastop, V. F. (1981). Coevolution of plants and insects. In "The Coevolving Biosphere" (P. L. Forey, ed.), pp. 179–190. British Museum (Natural History) and Cambridge University Press, Cambridge.

Faye, M., Rancillac, M. and David, A. (1980). Determinism of the mycorrhizogenic root formation in *Pinus pinaster* Sol. *New Phytol.* **87,** 557–565.

Flor, H. H. (1955). Host–parasite interaction in flax rust—its genetics and other implications. *Phytopathology* **45,** 680–685.

Francke-Grossmann, H. (1956). Grundlagen der Symbiose bei pilzzuchtenden Holzinsecten. *Verh. dt. zool. Ges.* **1956,** 112–118.

Futuyma, D. J. and Slatkin, M. (1983). Introduction. In "Coevolution" (D. J. Futuyma and M. Slatkin, eds), pp. 1–13. Sinauer Associates, Sunderland, Mass.

Garling, L. (1979). Origin of ant–fungus mutualism: a new hypothesis. *Biotropica* **11,** 284–291.

Gianinazzi-Pearson, V. and Gianinazzi, S. (eds) (1986). "Physiological and Genetical aspects of Mycorrhizae". Institut National du Recherches Agronomiques, Paris.

Gilbert, L. E. (1983). Coevolution and mimicry. In "Coevolution" (D. J. Futuyma and M. Slatkin, eds), pp. 263–281. Sinauer Associates, Sunderland, Mass.

Gilbertson, R. L. (1984). Relationships between insects and wood-rotting basidiomycetes. In "Fungus–Insect Relationships" (Q. Wheeler and M. Blackwell, eds), pp. 130–165. Columbia University Press, New York.

Greenwood, P. H. (1981). Coexistence and coevolution. Introduction. In "The Evolving Biosphere" (P. L. Forey, ed.), pp. 97–101. British Museum (Natural History) and Cambridge University Press, Cambridge.

Harborne, J. B. (1982). "Introduction to Ecological Biochemistry". Academic Press, London.

Harley, J. L. and Smith, S. E. (1984). "Mycorrhizal Symbioses". Academic Press, London.

Hawksworth, D. L. (1981). A survey of the fungicolous conidial fungi. In "Biology of Conidial Fungi" (G. T. Cole and [W.] B. Kendrick, eds), vol. 1, pp. 171–244. Academic Press, New York.

Hawksworth, D. L. (1988). The variety of fungal-algal symbioses, their evolutionary significance, and the nature of lichens. *Bot. J. Linn. Soc.* **96,** 3–20.

Heath, M. C. (1986). Fundamental questions related to plant–fungal interactions: can recombinant DNA technology provide the answers. In "Biology and Molecular Biology of Plant–Pathogen Interactions" (J. A. Bailey, ed.), pp. 15–27. Springer-Verlag, Berlin.

Hedberg, I. (ed.) (1979). Parasites as plant taxonomists. *Symb. bot. upsal.* **22**(4), 1–221.

Herrera, C. M. (1985). Determinants of plant–animal coevolution: the case of mutualistic dispersal of seeds by vertebrates. *Oikos* **44,** 132–141.

Hooper, D. J. and Stone, A. R. (1981). Role of wild plants and weeds in the ecology of plant parasitic nematodes. In "Pests, Pathogens and Vegetation" (J. M. Thresh, ed.), pp. 199–215. Pitman, Boston.

Humphries, C. J., Cox, J. M., and Nielsen, E. S. (1986). *Nothofagus* and its parasites: a cladistic approach to coevolution. In "Coevolution and Systematics" (A. R. Stone and D. L. Hawksworth, eds), pp. 55–76. Clarendon Press, Oxford.

Huxley, C. R. (1978). The ant-plants *Myrmecodia* and *Hydnophytum* (Rubiaceae), and the relationships between their morphology, ant occupants, physiology and ecology. *New Phytol.* **80,** 231–268.

Huxley, C. R. (1982). Ant-epiphytes of Australia. In "Ant–Plant Interactions in Australia" (R. C. Buckley, ed.), pp. 63–73. W. Junk, The Hague.

Jaenike, J. (1985). Genetic and environmental determinants of food preferences in *Drosophila tripunctata. Evolution* **39,** 362–369.

Jaenike, J., Grimaldi, D. A., Sluder, A. E. and Greenleaf, A. L. (1983). α-Amanitin tolerance in mycophagous *Drosophila. Science, N.Y.* **221,** 165–167.

Janzen, D. H. (1977). Why fruits rot, seeds mold, and meat spoils. *Am. Nat.* **111,** 691–713.

Janzen, D. H. (1979). Why food rots. *Nat. Hist.* **88,** 60–64.

Janzen, D. H. (1980). When is it coevolution? *Evolution* **34,** 611–612.

Janzen, D. H. (1981). The defenses of legumes against herbivores. In "Advances in Legume Systematics" (R. M. Polhill and P. H. Raven, eds), vol. 2, pp. 951–977. Royal Botanic Gardens, Kew.

Janzen, D. H. (1985). The natural history of mutualisms. In "The Biology of Mutualism" (D. H. Boucher, ed.), pp. 40–99. Croom Helm, London.

Jennersten, O. (1983). Butterfly visitors as vectors of *Ustilago violacea* spores between caryophyllaceous plants. *Oikos* **40,** 125–130.

Jones, C. T. (1984). Microorganisms as mediators of plant resource exploitation by insect herbivores. In "A New Ecology: Novel approaches to integrative systems" (P. W. Price, C. N. Slobodchikoff and W. G. Jaud, eds), pp. 53–99. Wiley, New York.

Kohlmeyer, J. and Hawkes, M. W. (1983). A suspected case of mycophycobiosis between *Mycosphaerella apophlaeae* (ascomycetes) and *Apophlaea* (Rhodophyta). *J. Phycol.* **19,** 257–260.

Kohlmeyer, J. and Kohlmeyer, E. (1974). Distribution of *Epichloë typhina* (ascomycetes) and its parasitic fly. *Mycologia* **65,** 77–86.

Kohlmeyer, J. and Kohlmeyer, E. (1979). "Marine Mycology. The Higher Fungi". Academic Press, New York.

Law, R. (1985). Evolution in a mutualistic environment. In "The Biology of Mutualism" (D. H. Boucher, ed.), pp. 145–170. Croom Helm, London.

Law, R. and Lewis, D. H. (1983). Biotic environments and the maintenance of sex—some evidence from mutualistic symbioses. *Biol. J. Linn. Soc.* **20,** 249–276.

Lemke, P. A. (1979). Coevolution of fungi and their viruses. In "Fungal Viruses" (H. P. Molitoris, M. Hollings, and H. A. Wood, eds), pp. 2–7. Springer-Verlag, Berlin.

Leppik, E. E. (1974). Evolutionary interactions between rhododendrons, pollinating insects and rust fungi. *Q. Bull. Am. Rhododendron Soc.* **28,** 70–89.

Leppik, E. E. (1975). Co-evolution of rust fungi, their hosts and obliquely adapted insects. [Mimeographed.] Botanical, Entomological and Phytopathological Society Meeting, Saskatoon, 18–22 August 1975.

Levin, D. A. (1976). The chemical defenses of plants to pathogens and herbivores. *A. Rev. Ecol. Syst.* **7,** 121–159.

Lewis, D. H. (1985). Symbiosis and mutualism: crisp concepts and soggy semantics. In "The Biology of Mutualism" (D. H. Boucher, ed.), pp. 29–39, Croom-Helm, London.

Lindeberg, G. (1986). Physiology of interactions between mycorrhizae and rhizosphere organisms. In "Physiological and Genetic Aspects of Mycorrhizae" (V. Gianinazzi-Pearson and S. Gianinazzi, eds), pp. 197–206. Institut National du Recherches Agronomiques, Paris.

Lovelock, J. (1986). Gaia: the world as living organism. *New Scientist* **112** (1539), 25–29.

Lumsden, R. D. (1981). Ecology of mycoparasitism. In "The Fungal Community" (D. T. Wicklow and G. C. Carroll, eds), pp. 295–318. Marcel Dekker, New York.

Madelin, M. F. (1968). Fungi parasitic on other fungi and lichens. In "The Fungi" (G. C. Ainsworth and A. S. Sussman, eds), vol. 3, pp. 253–269, Academic Press, New York.

Mamaev, B. M. (1975). "Evolution of Gall-Forming Insects—Gall Midges". (Transl. A. Crozy). British Library, Boston Spa.

Manocha, M. S. (1987). Cellular and molecular aspects of fungal host–mycoparasite interaction. *Z. PflKrankh. PflSchutz* **94,** 431–444.

Manocha, M. S., Balasubramanian, R. and Enskat, S. (1986). Attachment of mycoparasite with host but not with nonhost *Mortierella* species. In "Biology and Molecular Biology of Plant–Pathogen Interactions" (J. Bailey, ed.) pp. 59–69. Springer-Verlag, Berlin.

Martin, M. (1984). The role of ingested enzymes in the digestive processes of insects. In "Invertebrate–Microbial Interactions" (J. M. Anderson, A. D. M. Rayner, and D. W. H. Walton, eds), pp. 155–172, Cambridge University Press, Cambridge.

Meyer, J. and Dehne, H. W. (1986). The influence of VA mycorrhizae on biotrophic leaf pathogens. In "Physiological and Genetical Aspects of Mycorrhizae" (V. Gianinazzi-Pearson and S. Gianinazzi, eds), pp. 781–791. Institut National du Recherches Argonomiques, Paris.

Malloch, D. W. (1987). The evolution of mycorrhizae. *Can. J. Pl. Path.*, **9**, 398–402.

Moss, M. O. (1984). The mycelial habit and secondary product production. In "The Ecology and Physiology of the Fungal Mycelium" (D. H. Jennings and A. D. M. Rayner, eds), pp. 127–142. Cambridge University Press, Cambridge.

O'Connor, B. M. (1984). Acarine–fungal relationships: the evolution of symbiotic associations. In "Fungus–Insect Relationships" (Q. Wheeler and M. Blackwell, eds), pp. 354–381. Columbia University Press, New York.

Parlevliet, J. E. (1979). The co-evolution of host–parasite systems. *Symb. bot. upsal.* **22** (4), 39–45.

Parlevliet, J. E. (1986). Coevolution of host resistance and pathogen virulence: possible implications for taxonomy. In "Coevolution and Systematics" (A. R. Stone and D. L. Hawksworth, eds), pp. 19–34. Clarendon Press, Oxford.

Patelle, M. (1982). More mutualisms between consumers and plants. *Oikos* **38,** 125–127.

Person, C. (1968). Genetical adjustment of fungi to their environment. In "The Fungi" (G. C. Ainsworth and A. S. Sussman, eds), vol. 3, pp. 395–415. Academic Press, New York.

Petrini, O. (1984). Endophytic fungi in British Ericaceae: a preliminary study. *Trans. Br. mycol. Soc.* **83,** 510–512.

Petrini, O. (1985). Wirtsspezifität endophytischen Pilze bei einheimischen Eicaceae. *Bot. Helv.* **95,** 213–238.

Philipson, M. N. and Christey, M. C. (1985). An epiphytic/endophytic fungal associate of *Dathonia spicata* transmitted through the embryo sac. *Bot. Gaz.* **146,** 70–81.

Price, P. W. (1980). "Evolutionary Biology of Parasites". Princeton University Press, Princeton.

Price, P. W., Westoby, M., Rice, B., Atsatt, P. R., Fritz, R. S., Thompson, J. N., and Mobley, K. (1986a). Parasite mediation in ecological interactions. *A. Rev. ecol. Syst.* **17,** 487–505.

Price, P. W., Waring, G. L., and Fernandez, G. W. (1986b). Hypotheses on the adaptive nature of galls. *Proc. ent. Soc. Wash.* **88,** 361–363.

Rabotnov, T. A. (1980). On some problems of the coevolution of organisms. *Phytocoenologia* **7,** 1–7.

Rawlins, J. E. (1984). Mycophagy in Lepidoptera. In "Fungus–Insect Relationships" (Q. Wheeler and M. Blackwell, eds), pp. 382–423. Columbia University Press, New York.

Rayner, A. D. M., Brasier, C. M., and Moore, D. (eds) (1987). "Evolutionary Biology of the Fungi". Cambridge University Press, Cambridge.

Reichman, O. J. and Rebar, C. (1985). Seed preferences by desert rodents based on levels of mouldiness. *Anim. Behav.* **33,** 726–729.

Reichman, O. J., Wicklow, D. T., and Rebar, C. (1985). Ecological and mycological characteristics of caches in the mounds of *Dipodomys spectabilis*. *J. Mammal.* **66,** 643–651.

Reichman, O. J., Fattaey, A., and Fattaey, K. (1986). Management of sterile and mouldy seeds by a desert rodent. *Anim. Behav.* **34,** 221–225.

Rice, B. and Westoby, M. (1982). Heteroecious rusts as agents of interference competition. *Evol. Theor.* **6,** 43–52.

Rosenthal, G. A. (1983). A seed-eating beetle's adaptations to a poisonous seed. *Scient. Am.*, Nov. 1983, 164–171.

Sagan, L. (1967). On the origin of mitosing cells. *J. theor. Biol.* **14,** 225–275.

Sagara, N. (1981). Occurrence of *Laccaria proxima* in the grave site of cat. *Trans. mycol. Soc. Japan* **22,** 271–275.

Savile, D. B. O. (1954). Taxonomy, phylogeny, host relationship and phytogeography of the microcyclic rusts of Saxifragaceae. *Can. J. Bot.* **32,** 400–425.

Savile, D. B. O. (1976). Evolution of the rust fungi (Uredinales) as reflected by their ecological problems. In "Evolutionary Biology" (M. K. Hecht, W. C. Steere, and B. Wallace, eds), vol. 9, pp. 137–207. Plenum Publishing, New York.

Savile, D. B. O. (1979). Fungi as aids to higher plant classification. *Bot. Rev.* **45,** 377–503.

Siegel, M. R., Latch, G. C. M., and Johnson, M. C. (1987). Fungal endophytes of grasses. *A. Rev. Phytopath.* **25,** 293–315.

Siegel, R. K. (1979). Natural animal addictions: an ethological perspective. In "Psychopathology in Animals" (J. D. Keehn, ed.), pp. 29–60. Academic Press, New York.

Smith, D. C. and Douglas, A. E. (1987). "The Biology of Symbiosis". Edward Arnold, London.

Southwood, T. R. E. (1985). Interactions of plants and animals: patterns and processes. *Oikos* **44,** 5–11.

Starmer, W. T. (1982). Associations and interactions among yeasts, *Drosophila* and their habitats. In "Ecological Genetics and Evolution; the cactus–yeast–drosophila model system" (J. S. F. Barker and W. T. Starmer, eds), pp. 159–174. Academic Press, Sydney.

Starr, M. P. (1975). A generalized scheme for classifying organismic associations. *Symp. Soc. exp. Biol.* **29,** 1–20.

Stone, A. R. and Hawksworth, D. L. (eds) (1986). "Coevolution and Systematics". Clarendon Press, Oxford.

Swift, M. J. and Boddy, L. (1984). Animal–microbial interactions in wood decomposition. In "Invertebrate–Microbial Interactions" (J. M. Anderson, A. D. M. Rayner, and D. W. H. Walton, eds), pp. 89–131. Cambridge University Press, Cambridge.

Taylor, F. J. R. (1983). Some eco-evolutionary aspects of intracellular symbiosis. *Int. Rev. Cytol., Suppl.* **14,** 1–28.

Thompson, J. N. (1982). "Interaction and Coevolution". Wiley-Interscience, New York.

Thompson, J. N. (1986). Patterns in coevolution. In "Coevolution and Systematics" (A. R. Stone and D. L. Hawksworth, eds), pp. 119–143. Clarendon Press, Oxford.

Tinker, P. B. (1984). The role of microorganisms in mediating and facilitating the uptake of plant nutrients from soil. *Pl. Soil* **76,** 77–91.

Trappe, J. M. and Molina, R. (1986). Taxonomy and genetics of mycorrhizal fungi: their interactions and relevance. In "Physiological and Genetical Aspects of Mycorrhizae" (V. Gianinazzi-Pearson and S. Gianinazzi, eds), pp. 133–146. Institut National du Recherches Agronomiques, Paris.

Van Valen, L. M. (1973). A new evolutionary law. *Evol. Theor.* **1,** 1–30.

Van Valen, L. M. (1983). How pervasive is coevolution? In "Coevolution" (M. H. Nitecki, ed.), pp. 1–19. University of Chicago Press, Chicago.

Verkaar, H. J. (1986). When does grazing benefit plants? *Trends Ecol. Evol.* **1**, 168–169.

Verma, D. P. S. and Nadler, K. (1984). Legume-*Rhizobium* symbiosis: host's point of view. In "Genes involved in Microbe–Plant Interactions" (D. P. S. Verma and Th. Hohn, eds), pp. 57–93. Springer-Verlag, Vienna.

Wahl, I., Anikster, Y., Manisterski, J., and Segal, A. (1984). Evolution at the center of origin. In "The Cereal Rusts" (W. R. Bushnell and A. P. Roelfs, eds), vol. 1, pp. 39–77. Academic Press, Orlando.

Wheeler, Q. and Blackwell, M. (eds) (1984). "Fungus–Insect Relationships". Columbia University Press, New York.

Wicklow, T. D. and Carroll, G. C. (eds) (1981). "The Fungal Community". Marcel Dekker, Basel and New York.

Wicklow, T. D. and Rebar, C. (1988). Mold inoculum from cheek pouches of a granivorous desert rodent, *Dipodomys spectabilis*. *Mycologia*, in press.

Williams, A. G. and Orpin C. G. (1987). Polysaccharide-degrading enzymes formed by three species of anaerobic rumen fungi grown on a range of carbohydrate substrates. *Can J. Microbiol.* **33,** 418–426.

is obviously a downy mildew association. It is at least feasible that characteristic deformities due to fungal attack may have been preserved (Kevan *et al.*, 1975), but in fossils there may be insufficient material for a deformed state or disease syndrome to be recognized. Disease syndromes most likely to be associated with downy mildews (hyperplasia, hypertrophy or dwarfing) have not been identified. (Indeed, with extant fungi of any class there has been little emphasis on the coevolutionary aspects of the taxonomic distribution of disease syndromes, although they exist: consider wilt diseases and anthracnoses in relation to the diagram in Dick and Hawksworth, 1985.) Preservations and pre-fossil records usually relate to oospores or oogonia (Wolf, 1966, 1968). However, there must be considerable doubt regarding the ability to differentiate unknown fossil oomycete oogonia from unknown fossil zygospores of Zygomycotina such as the Endogonales or Zoopagales. Figures 1 and 2 illustrate only one aspect of the problem. Other resemblances can be envisaged, such as a comparison between an oogonium without antheridia and the zygosporangium such as that of *Dimargaris*. An indisputably oomycete fossil would need to have additional features beyond the thick-walled, spherical shape. For the Oomycetes it is probably unlikely that the fossil evidence will ever be adequate to support proposals of coevolution.

Phylogeny

An outline phylogeny of the heterokont fungi (based on Dick *et al.*, 1984; Dick, 1988) is given in Figure 3. However, a major systematic reorganization of the endobiotic biflagellate and reputedly heterokont fungi is required (Dick, in preparation).

In contrast to earlier suggestions (Pirozynski, 1976), the consensus of opinion (International Society for Evolutionary Protistology, 7th Meeting, 1987) is that the Chromista (which includes chromophyte algae and heterokont fungi) is of relatively recent origin amongst Protoctista. It may even post-date the origin of true fungi.

Biochemistry

When a higher taxon has very few representatives, and when some or all are saprophytic, coevolutionary links are difficult to establish. Discussion of coevolution becomes restricted to the few genera of Oomycetes cited in Figure 3. A biochemical approach to coevolution is also difficult to evaluate because most of the more highly evolved parasites have yet to be cultured outside their hosts. Only a few species have been grown in biphasic systems using host tissue culture (*Bremia* (1), *Peronospora* (3), *Plasmopara* (1), *Pseudoperonospora* (1), *Sclerophthora* (1), and *Sclerospora* (3); Ingram, 1981). Parallels have to be

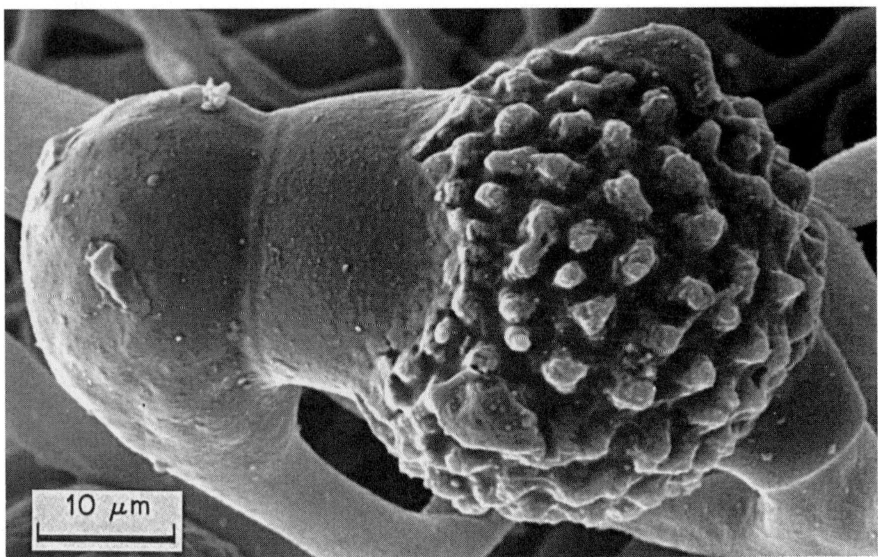

Figure 1. Scanning electron micrograph of *Zygorhynchus psychrophilus* zygosporangium.

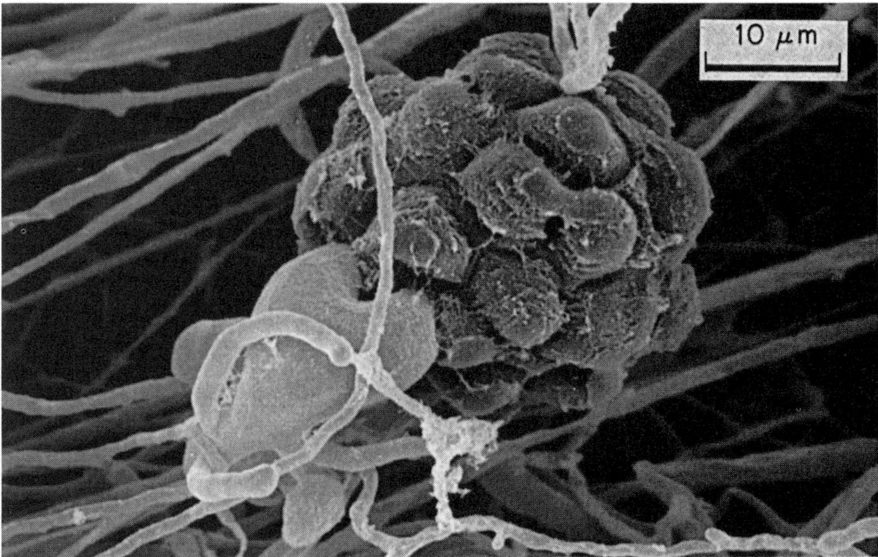

Figure 2. SEM of *Pachymetra chaunorhiza* oogonium and antheridium.

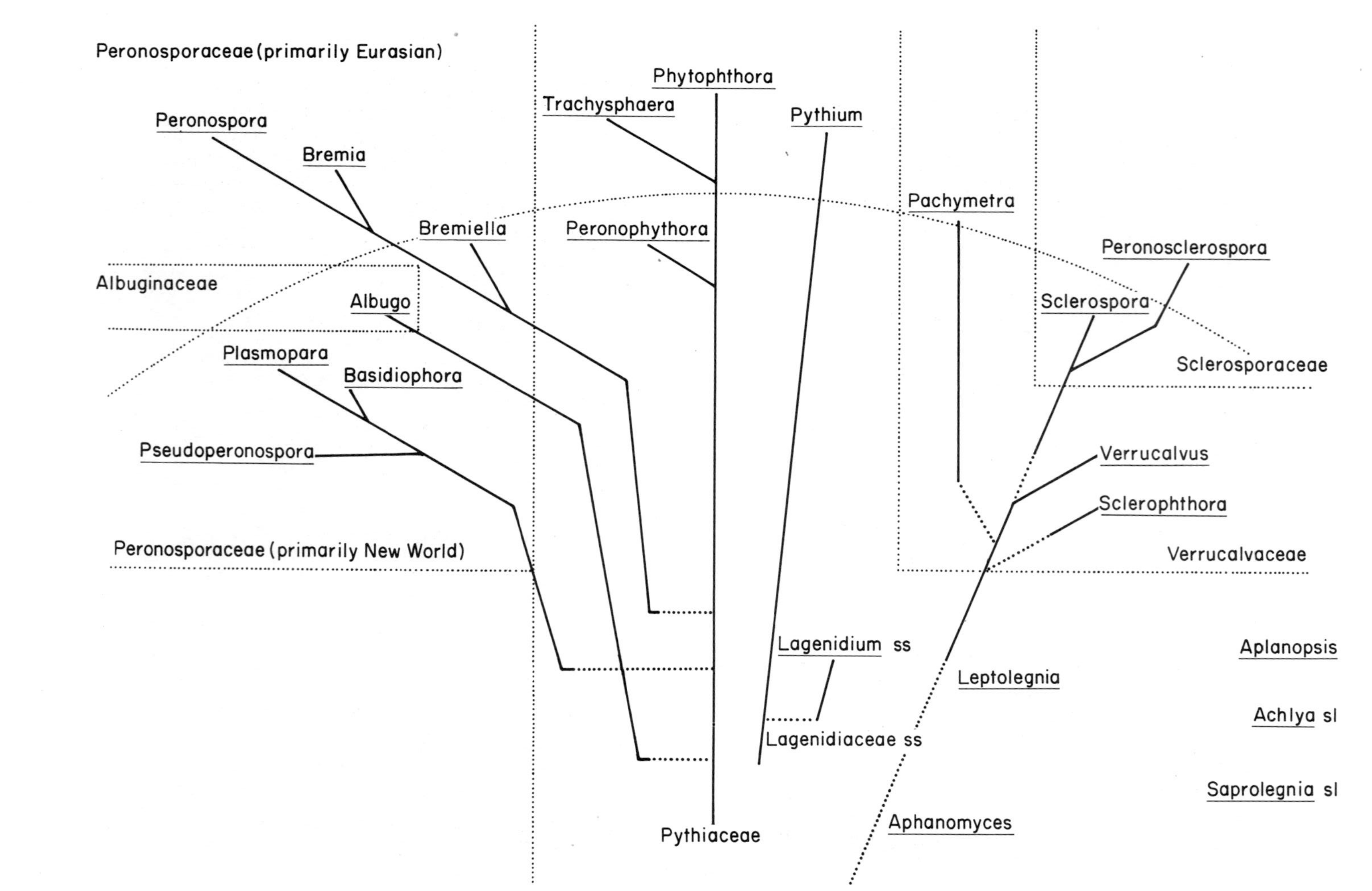
Peronosporaceae (primarily Eurasian)
Phytophthora
Trachysphaera
Pythium
Peronospora
Bremia
Pachymetra
Bremiella
Peronophythora
Peronosclerospora
Albuginaceae
Albugo
Sclerospora
Plasmopara
Sclerosporaceae
Basidiophora
Pseudoperonospora
Verrucalvus
Sclerophthora
Peronosporaceae (primarily New World)
Verrucalvaceae
Lagenidium ss
Aplanopsis
Leptolegnia
Achlya sl
Lagenidiaceae ss
Saprolegnia sl
Aphanomyces
Pythiaceae

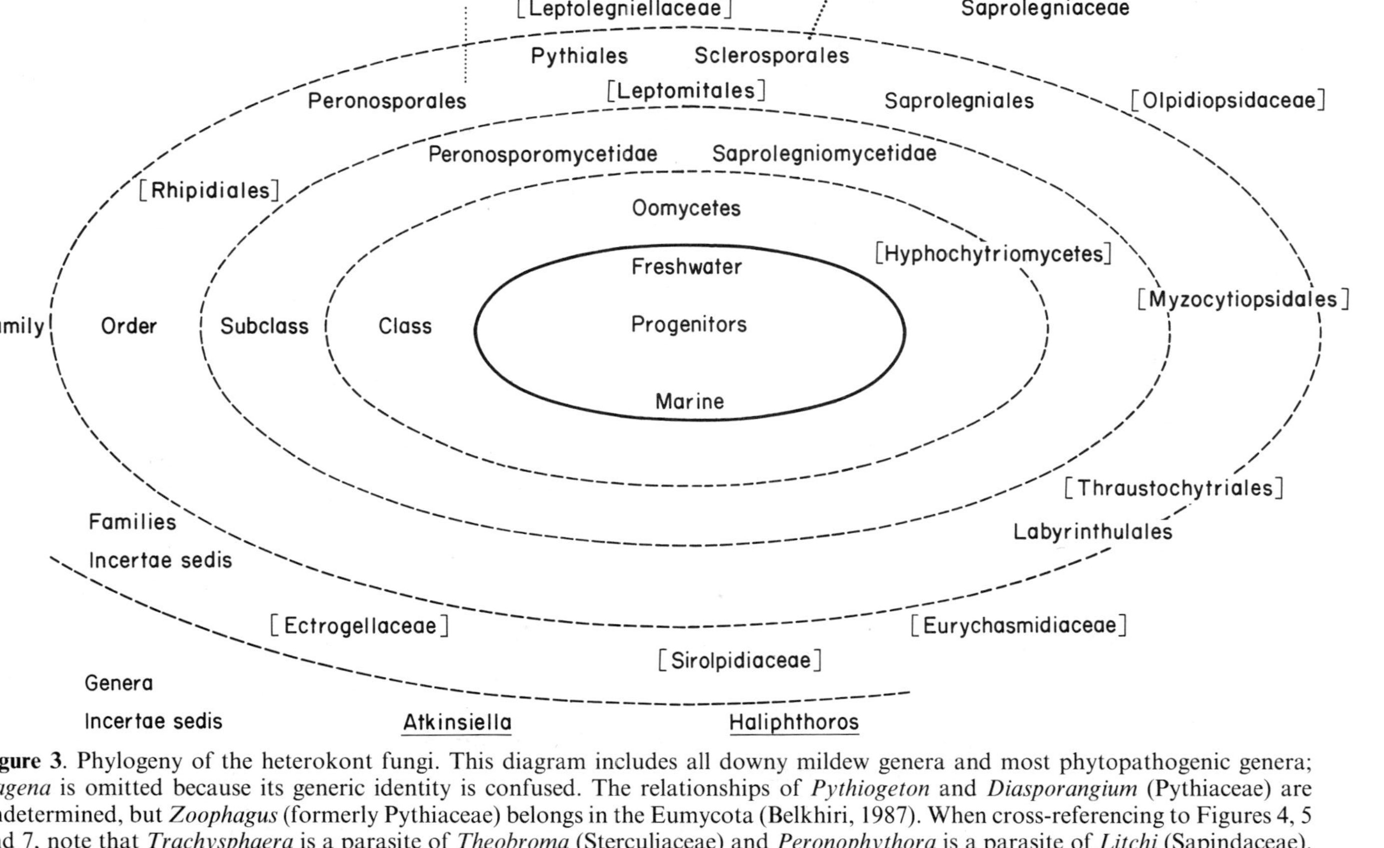

Figure 3. Phylogeny of the heterokont fungi. This diagram includes all downy mildew genera and most phytopathogenic genera; *Lagena* is omitted because its generic identity is confused. The relationships of *Pythiogeton* and *Diasporangium* (Pythiaceae) are undetermined, but *Zoophagus* (formerly Pythiaceae) belongs in the Eumycota (Belkhiri, 1987). When cross-referencing to Figures 4, 5 and 7, note that *Trachysphaera* is a parasite of *Theobroma* (Sterculiaceae) and *Peronophythora* is a parasite of *Litchi* (Sapindaceae). No genera are listed for the higher taxa in brackets. The zone between the Peronosporomycetidae and Saprolegniomycetidae (Leptolegniellaceae, Sclerosporales, and Leptomitales) is a "grey area" of affinity. Some taxa are listed as 'ss' *sensu stricto* or 'sl' *sensu lato*. The dotted arc indicates the approximate boundary between zoosporic and non-zoosporic taxa. The dotted straight lines indicate the boundaries of families of downy mildews. This diagram supersedes that in Dick (1988).

drawn from data for facultative parasites that are thought to be related to the obligate parasites.

THE EUKARYOTE

Eukaryotes, with mitochondria and plastids are "composite" organisms. Studies are still awaited that may resolve the frequency of cell melding over evolutionary time; meldings that have resulted in phylogenetic lines of composite organisms (e.g., Morrall and Greenwood, 1982). Plastids (some perhaps now achlorophyllous), mitochondria, and even nuclei, all from diverse sources, may have contributed to the unit now recognized as the heterokont eukaryote cell.

Two organelles are particularly characteristic of heterokont organisms. These are the heterokont flagella and the mitochondria with tubular cristae. Heterokont flagellation is characterized by the possession of a tinsel flagellum that is about ten times more powerful as a "motor" than the whiplash flagellum (Holwill, 1982). The energy demands of the tinsel flagellum will be correspondingly greater and these demands may depend upon mitochondrial efficiency as well as the cytoplasmic fuel source. The need for high levels of oxygen in the environment during the motile phase is well known (Carlile, 1983).

Conventionally, it may be assumed that the mitochondrion (prokaryote bacterioid) was incorporated at the same time as, or may even have been an essential adjunct to, heterokont flagellation. The evolution of the tinsel flagellum not only affected motile efficiency, it also affected sensory perception, in that the anterior part of the cell was no longer the plasma membrane of the cell proper, but the plasma membrane at the tip of the tinsel flagellum. In the brown algae, sensory and attachment functions are located in this flagellum (Müller and Falk, 1973).

This scenario does not exclude the hypothesis that subsequent "consumptions" of heterokont by heterokont may have occurred, so that the organisms recognized today may be stable composites of heterotroph and heterotroph or heterotroph and ex-autotroph. In this connection, the results of McNabb *et al.* (1987) are tantalizing in that the mitochondrial DNA of *Pythium* shows an inverted repeat of a length (27–29 kilobase pairs) that is quite different from that in eumycote mitochondria (4–5 kb) or *Achlya* (10 kb) but very similar to that in chloroplasts (20–28 kb) (Whitfield and Bottomley, 1983).

The possibility of such reticulate evolution at the cellular level is the reason that there has been no attempt to make linear connections in Figure 3 between the groups of heterokont fungi. The obvious vehicles for the evolution of composite organisms are the naked flagellate cell in the asexual phase and the gametangial fusion with juxtaposition of naked protoplasts in the sexual phase. If wholesale transfer of organelles is not involved, genetic transfer may have been mediated by plasmids.

It must not be assumed that such amalgamations would have been confined to closely related organisms. What roles might have been played by endobiotic protozoa, plasmodiophoromycetes, or olpidiopsoid fungi? All have naked protoplasts that can be totally enclosed in a vesicle of the host cell.

From what little is known of the construction, biochemistry, and storage functions of vesicle systems, including dense body vesicles (DBVs) and microbodies (Powell *et al.*, 1985), the combination of mitochondria, flagellation, etc., may have evolved from a reticulate relationship amongst coevolutionarily-linked organelles on more than one occasion in evolutionary time.

The chemical identity of enzymes performing similar functions for different organelles, and the difference, if any, in the DNA base sequence that codes for these organelles in related species, have still to be established. It is possible that comparative studies will clarify the relatedness of different organelles, and the conserved nature of certain forms and functions.

The starting point for coevolution of organisms may have been the coevolution of the organelles and biochemical pathways within the eukaryote cell.

ENVIRONMENTAL OPPORTUNITIES

There must be, or have been, proximity between populations of the taxa involved for coevolution to have taken place. Coevolution may embrace more than two taxa and the populations need not be large, stable, or of equivalent size. However, if the potential parasite is sexually reproducing, there is probably a minimum population size below which a coevolutionary relationship is unlikely to develop. This postulate could be argued on the out-breeding and colonization potentials of the parasite as well as the ability of the host population to develop resistance. Proximity may be of long standing, with a gradually established, balanced dependence. Alternatively, it may happen suddenly, either when a vulnerable alien host becomes introduced to an environment containing a previously stabilized host–parasite relationship, or when the spore population of a parasite is alien but able to exploit a susceptible but previously inaccessible host (Hijwegen, this volume, Chapter 3).

There must be physical or chemical similarities or analogues that enable an appropriate degree of association between previously separated populations. Too great a vulnerability will lead to an unstable and ephemeral relationship. The essence of coevolution is adaptive change in balanced relationships. It is possible that chance associations may lead to new relationships, as has been proposed by Baum and Savile (1985) for certain rusts. This may be more possible for parasites that produce a limited mycelium and for which physical rather than chemical environmental factors are more important. Chance associations leading to coevolution must be less likely for fungi that are essentially systemic, because there would be less likelihood that either host or parasite would survive long enough to reach reproductive maturity.

Coevolution is most probable within the communities of an ecosystem and amongst biochemically-related hosts. Although populations must be involved, in order to establish a diversity in the gene pool, the ultimate origin of each coevolutionary relationship rests with individuals.

Interacting organisms must be able to come into contact, and there must be sufficient similarity for nutritional requirements to be satisfied. Frequently, this will be because new hosts are phylogenetically close to former hosts. But phylogenetically distant or unrelated potential hosts may also provide the critical factors for the nutritional environment of the parasite. The factors involved are not necessarily the most evident and the major nutritional requirements, carbon and nitrogen sources, are unlikely to be crucial.

For some relationships, hydrogen-ion concentration may be critical, while for others there may be an interaction between adjacent plasma membranes that will lead to suitable nutrient flow. In other relationships, there may be the provision of an intermediate metabolite for the parasite, or the removal of a waste product of the heterotroph that provides the essential environmental condition for successful establishment of a host–parasite relationship.

Evidence for coevolution should not be sought exclusively within one ecosystem, or even one community, nor within one phylogenetic line. On the other hand, one must doubt the tacit acceptance (e.g., Karling, 1981; Sparrow, 1960) that obligately parasitic species with limited host ranges can be both closely related and have diverse hosts in widely disparate ecosystems. There would have to be very special reasons for allowing coevolutionary, phylogenetic unity between one parasite restricted to an order of freshwater algae and another restricted to crabs' eggs, as is permitted without apparent question in *Lagenidium*. Coevolutionary considerations may clarify and correct the taxonomic precepts (Dick, in preparation).

ORGANISMAL ASSOCIATIONS

Kinds of association

For the heterokont fungi, four levels of organism–organism association can be recognized.

Necrotrophic parasitism

In necrotrophic parasitism, the fungus can grow and reproduce without the contemporaneous presence of other living protoplasm. Necrotrophs are sometimes facultative saprophytes. Such fungi are frequently of sporadic occurrence. They have an aggressive and destructive mode of attack and colonization, often involving rapid growth rates and powerful extracellular enzymes. *Pythium* species are obvious examples, but associations involving *Pythium* species are not always, or necessarily, indiscriminate. Some nutrient sources are preferred by certain species. In nature, some sources may be exclusively used (e.g., fungicolous *Pythium* species; Deacon, 1976), and Hardman (1985) suggested that saprophytic *Pythium* species show different responses to the presence of grass or *Hebe* baits; the former is more effective for *Pythium*, and the latter is recommended for *Phytophthora*. *Phytophthora* also possesses a range of degrees of specialization, although, in contrast to *Pythium*, the associations are usually with dicotyledons (*P. palmivora* is a notable exception). Some species have very wide host ranges (e.g., *P. cinnamomi* on *c*. 1000 hosts; Zentmyer, 1980) while others are usually confined to one angiosperm order (e.g., *P. infestans* on Polemoniales), to a particular species (e.g., *P. colocasiae* on *Colocasia esculenta*), or to a particular part of a species (e.g., *P. megakarya* on *Theobroma cacao* pods).

Parasitoidal associations

Parasitoidal associations are analogous to predation. As with predation and facultative parasitism, they are essentially necrotrophic but differ in being endobiotic and normally taxon-specific. Newell *et al.* (1977) refer some species of *Lagenidium* and *Myzocytium* to this mode, but in view of the doubts as to the position of these fungi, *Olpidiopsis* is a more appropriate heterokont fungal example (Slifkin, 1961; Martin and Miller, 1985). Unlike facultative parasitism, the initial contact between the interacting species is of overriding importance; it is likely to be biochemically mediated through host exudates. It is debatable whether all holocarpic and eucarpic endobionts (e.g., *Olpidiopsis varians* and *Aphanomyces parasiticus*, respectively, on *Aplanopsis*) should be regarded as parasitoidal associations. An important factor is the harmony between population sizes and generation times. Dick (1976) has suggested that stability between populations may be achieved with about 10% infection. More deviant ratios would result in erratic occurrences. In this respect, a predator–prey population relationship may be appropriate.

Biotrophic parasitism

Biotrophic parasitism is much more complex. The first evolutionary steps must have been the development of a restraint on cell penetration and a curb to the rapid growth rates shown by fungi such as *Pythium*. It could be envisaged that these changes developed as a result of genetic lesions rendering the fungus dependent upon some external source of an intermediate metabolite; dependence would lead to obligate parasitism. Secondary events would have been the elaboration of haustorial systems; an increase in the generation time of the fungus, and fungal stimulation of host cells to give hypertrophy and hyperplasia.

Not all downy mildews are fully biotrophic. Significant cell damage is caused, for example, by *Plasmopara viticola*. The host plasma membranes become excessively leaky, resulting in the distinctive greasy or wet appearance–a moderated manifestation of the symptom associated with wet rots caused by certain species of *Phytophthora* (Keen and Yoshikawa, 1983) and probably coming from a similar biochemical interaction. Other downy mildews are capable of developing systemic infections that require a greater tolerance on the part of the host. Clarke (1983) has drawn attention to the different concepts of parasitism and the causing of disease, but symptomless parasitism, which would be the ultimate in host tolerance, has not yet been reported for the hosts of oomycete parasites.

Special form relationships

From the coevolutionary viewpoint, there are distinctions to be drawn among obligate parasitism, species-specific parasitism, and special-form relationships. Whereas obligate parasitism merely requires the presence of a regular (but possibly periodic) and renewable (but possibly highly transient) nutrient availability from living protoplasm, species-specific parasitism implies a much more restricted tolerance range for potential complementary metabolisms. I prefer to view this as a tolerance range rather than as absolute chemical requirements. The latter would require an improbably large number of genetic lesions to explain race-specific parasitism between related parasites and related hosts.

The endpoint of this progression is the race concept of the special form for which biochemical compatibility is presumed to be the only apparent distinguishing feature. This may be merely the result of extremely narrow tolerance ranges for a number of factors. But it may be, as with race induction in response to resistance cultivar production, a gene-for-gene evolution that may function through a variety of physiological or morphological requirements. For the oomycetes (*Bremia*), a recent discussion on special forms is that by Skidmore and Ingram (1985).

The genetic specializations implicit in special forms make such taxa the ultimate lateral twigs of the phylogenetic tree. They are distant from the leader shoots of coevolution.

Nature of the associations

In the Oomycetes, many organismal associations of a very exacting nature have arisen. Factors that are important for individual associations, but which have lesser relevance to the broader theme of coevolution, are excluded here; these include competition, antagonism, taxis or synergism, mechanisms and morphology of penetration, colonization, and haustorial morphology.

More than two taxa may have roles in one coevolutionary example. This may be seen in examples of host, parasite, and vector, as in Dutch Elm Disease, or in heteroecious rusts, but other possible organismal complexes can be envisaged, for example, the suggestion (Sansome and Sansome, 1974) that *Albugo* and *Peronospora* are synergistic on crucifers, and the susceptibility of beans to *Pseudoperonospora cubensis* only after prior infection by *Uromyces* (Yarwood, 1977).

I would also suggest that host–root associations (mycorrhizas or weak root parasites) might so affect the wellbeing of the host as to make a significant change (either way) to its vulnerability to parasitism of aerial parts. In experimental studies of biological control (Elad and Chet, 1987), induced root communities have been explored within root systems. As far as I am aware, communities on and in roots have not been tested in connection with the metabolism of the whole angiosperm plant in different growing conditions or with downy mildew parasitism of the aerial shoot.

One interesting conjunction, which needs more substantiation, is that between ruderal herbs (e.g., Brassicaceae and Caryophyllaceae) and downy mildews. Ruderal herbs lack mycorrhizas (Gerdemann, 1968). It is interesting to speculate that perhaps the absence of mycorrhizal associations alters the flow of metabolites, and perhaps the manufacture of certain secondary metabolites, so that an accumulation of essential oils, alkaloids, and saponins is facilitated. For most organisms, including fungi, these substances are antagonistic or repellent. For a few organisms (Oomycetes?) they may be attractants, or serve as unexpected intermediates to essential metabolism.

There are two interconnected factors in this hypothesis: the coevolutionary reliance by the parasite on a host species, and the restrictive nature of this reliance to particular metabolitic pathways. These pathways, or the specific metabolites produced, may occur in organisms of different phylogeny, and they may only become manifest in certain populations because of environmental circumstances.

Incidence of associations

A statistical approach is inappropriate if the numbers of known downy mildews and the numbers of potential hosts are considered. There are, at most, only a few hundred known oomycete obligate parasites and a few hundred thousand potential plant hosts. Species-for-species correlations will not help. However, taking a sufficiently broad overview, emphasizing the host ordinal rather than family level (*contra* Palti and Kenneth, 1981), a possible pattern emerges. Three broad groups of dicotyledons are preferred hosts for the Peronosporaceae, Albuginaceae, and *Phytophthora*, and one group of monocotyledons is host to the Sclerosporaceae, Verrucalvaceae and *Pythium*. This disjunct distribution merits examination and explanation.

The most obvious feature of the distribution of the parasites of dicotyledons (Figure 4(a)–(d)) is that it is not primarily the more primitive or ancient orders of angiosperms that are affected. Despite the suggestion of an aquatic ancestral origin for the oomycetes, as evidenced by the production of zoospores, the host angiosperm orders are not noted for their aquatic members. The habitats of these hosts are diverse and an enormous morphological and anatomical range is encompassed. In this example, coevolution is not therefore specialization from closely associated ancestral stocks. It is a common fallacy to regard coevolution as necessarily an indication of phylogeny (e.g., Crute, 1981). The fact that the widest spectrum of hosts is shown by the morphologically "advanced" (i.e., lacking zoospores) genus *Peronospora* is not incongruous if the attraction is biochemical. Proximity to the aquatic habitat is less likely to lead to the comprehension of coevolution than physiology. For species-specific parasitism the more intensively studied phenomena of suppression and inhibition are less important than the elements of attraction and stimulation.

BIOCHEMICAL DEPENDENCE

If an organism is an obligate parasite that cannot be grown apart from its living host, either it requires particular chemicals that have not yet been identified or the organism is intolerant of arbitrary levels of fluctuations in the concentrations and rates of supply of nutrients. The efficiency of waste removal may be a contributory factor.

There are very few examples of data to support or refute either of these contentions. Extrapolations have to be made from studies of related fungi that can be grown axenically. Cantino (1955), Gleason (1976) and Hohl (1983) have reviewed nutritional aspects of the Peronosporomycetidae. Cantino's hypothesis that the loss of ability to utilize certain inorganic forms of nitrogen and sulphur represents irreversible genetic lesions is still accepted. Genetic lesions that lead to absolute requirements for particular chemicals are known to occur among races of a parasite (e.g., *Phytophthora infestans*; Hohl, 1983). There are no suggestions that requirements are invariably linked to host range restriction. Such changes must be regarded as part of the heterogeneity of the gene pool. Vitamin requirements are also well known (e.g., Hohl, 1983), but auxotrophs can usually survive on the low concentrations normally leaked from autotrophs, so these requirements are also unlikely to be a determining factor for obligate parasitism.

The problem with nutritional studies (cf., Hohl, 1983) is that they have to take place in artificial environments. Conditions conducive to maximum and prolonged vegetative growth may not be the most advantageous (i.e., fittest or most competitive) in the natural environment. Not even for *Phytophthora* is there any sound case for suggesting that particular nutrient requirements are essential prerequisites for successful attack. For an understanding of coevolution, it may be necessary to distinguish between nutritional regimes that give vegetative growth enhancement and those that relate growth to generation time and efficiency of reproduction.

If parasite dependence is not based on a demand for particular chemical units, might the dependence be based upon an "empathy" between certain crucial metabolic pathways of host and parasite, so that the catabolism and anabolism of both are in accord? Different host pathways may be pre-eminent for different parasites. Thus, individuals of a single host species may be infected by several parasites: these parasites may also encompass different degrees of host specificity. Most of the downy mildews are leaf and herbaceous stem parasites. Photosynthesis may therefore be an important physiological influence. Variations from the C_3 metabolism (C_4 and crassulacean acid metabolism, CAM) are common; they are assumed to have arisen independently in different angiosperm orders, but the mapping of their systematic distributions is still in its infancy (Moore, 1982). It is probable that both carbon intermediates and ionic balances are different in these systems. The availability of carbon to parasites may be different as well.

If the angiosperm ordinal distributions of C_4 and CAM are compared (and also compared with phytoalexin distribution, Figure 5), it can be seen that the only orders producing phytoalexins, but for which no C_4 or CAM has been reported, are the Fabales, Umbelliferales, Malvales and Urticales. The Caryophyllales and Euphorbiales, and probably the Asterales and Liliales, are orders in which C_4, CAM and phytoalexin production are common. The Poales have phytoalexins and the panicoid grasses in particular, are noted for C_4 metabolism and flavonoids. The relationships of the graminicolous Sclerosporales with C_4 grasses (phytoalexin producers?) are probably close. However, the association of C_4 orders with *Plasmopara* is poor, and it is non-

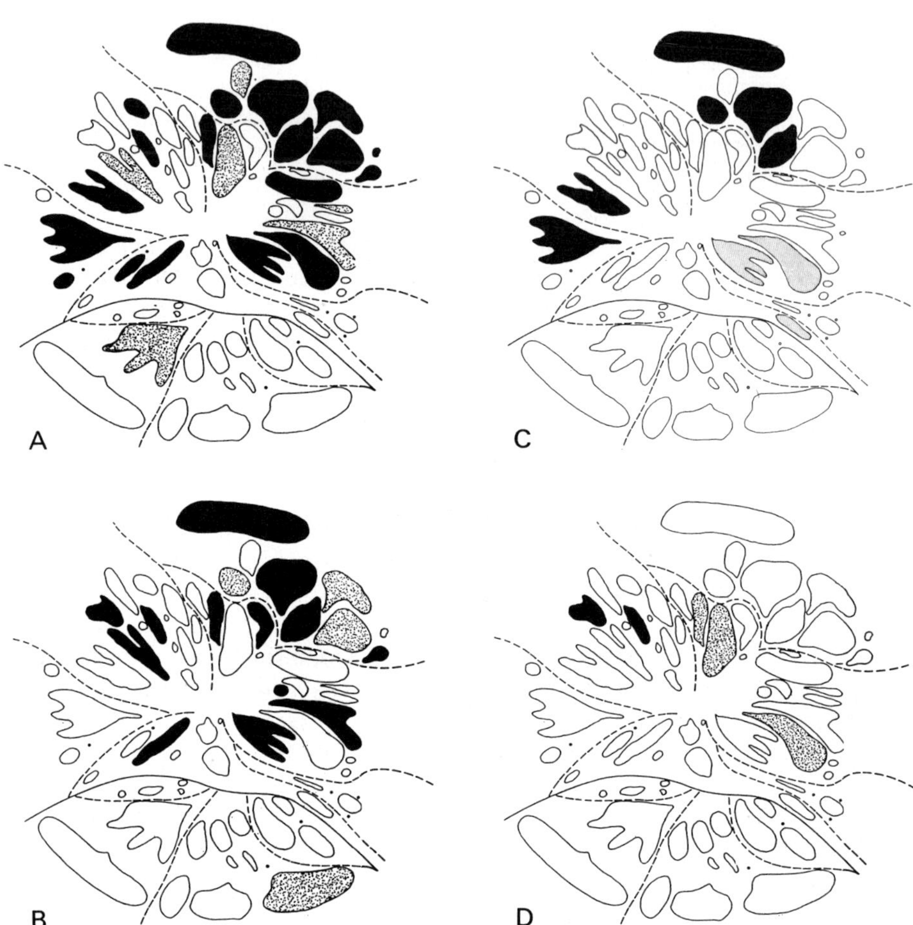

Figure 4. Occurrence of downy mildews on angiosperm hosts, grouped according to orders arranged following Stebbins (1974). For key to ordinal names see Figure 5D, and cross-reference to Figure 6 for relevant families. A, *Peronospora*; B, *Plasmopara*; C, *Albugo*; D, *Pseudoperonospora*. Black, principal hosts; stippled, minor or doubtful hosts. Compare Figures 4A and 4C with 5A; and Figures 4C and 4D with 5B.

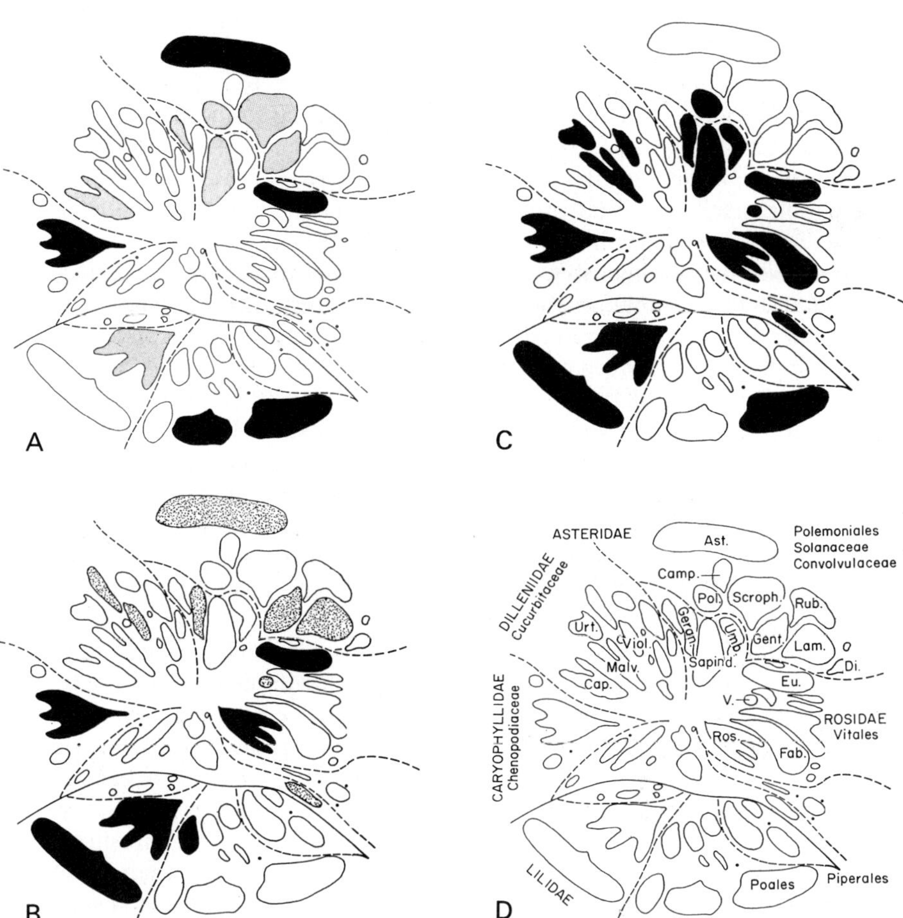

Figure 5. Occurrence of C_4, CAM photosynthetic pathways (Moore, 1982) and phytoalexins in angiosperm orders. A, C_4 plants; B, CAM plants; C, phytoalexins reported; D, key to supra-orders and orders of angiosperms. Black, present; stippled, few or doubtful records. *Abbreviations:* Cap, Capparales; Urt, Urticales; Malv, Malvales; Viol, Violales; Geran, Geraniales; Sapind, Sapindales; Umb, Umbellales; Pol, Polemoniales; Camp, Campanulales; Ast, Asterales; Scroph, Scrophulariales; Gent, Gentianales; Rub, Rubiales; Lam, Lamiales (Plantaginales distal spot); Di, Dipsacales; Eu, Euphorbiales; V, Rhamnales (Vitaceae); Fab, Fabales; Ros, Rosales.

Compare Figure 5A with 4A and 4B; and Figure 5B with 4B and 4D.

existent with *Pseudoperonospora*. For *Plasmopara* an association is possible with CAM in the Rosidiae and, similarly, for *Pseudoperonospora* and the Violales. The C_4 relationship has more potential with *Albugo* and *Peronospora*.

It would be unwise to attach too much significance to such ordinal comparisons, because it is still fortuitous that a particular angiosperm has been studied for photosynthesis, phytoalexin production, and susceptibility to Peronosporomycetidae. But do these comparisons hint at the existence of "metabolic packages" that predispose certain angiosperm taxa to coevolutionary relationships with the Oomycetes?

Whatever the biochemistry underlying attraction to a particular host, and stimulation to germination and colonization by the parasite, there are well-documented examples of parasite-mediated modification of host physiology after establishment. Green ear hyperplasia of pearl millet caused by *Sclerophthora* (Williams, 1984), hypoplasia of sunflower by *Plasmopara* (Sackston, 1981), and the well-known hypertrophy of crucifer stems by *Albugo* are three of the clearest examples relating to growth substances. The precise mechanisms of the biochemical modifications have not been researched.

A comparative survey of the occurrence of growth-substance-related disease symptoms in facultative as well as obligate parasites is also required, for there could be a phylogenetic pattern in the distributions of these symptoms.

Galston (1983) linked polyamine metabolism with the metabolism of growth substances such as gibberellins and cytokinins. While most attention has been given to secondary metabolites such as flavonoids and alkaloids, he regards polyamines as neglected but probably important secondary metabolites. Galston notes that the polyamine putrescine is probably to be found in all cells. (Could this be the cause of the characteristic odour of rot caused by *Phytophthora*?) Other polyamines may help to control the nuclear cycle. Lysine is a precursor for many polyamines. Unlike most fungal parasite–host systems, both angiosperm host and oomycete parasite have the same (DAP) lysine synthesis pathway. Vaziri-Tehrani and Dick (1980) have suggested that this pathway, which requires fewer enzymes than the AAA pathway and therefore may be more economical, could account for the relatively high levels of lysine in the proteins of oomycete cell walls. The second active amino group of lysine could also aid nutrient uptake. Galston also reported a link between CAM and polyamines mediating cellular pH stasis. Perhaps the polyamines could be another component of the "metabolic package"?

The distinction between the coevolution of parasite and host, and the predisposition of one host individual to parasitism, must be made. The importance of secondary metabolites to the establishment of a successful parasitic relationship must be considered. Genotypic differences in the host production of secondary metabolites may contribute to the resistance or susceptibility of the individual or cultivar. This has been shown by Walker (Tarr, 1972, p. 251) in relation to catechol production by *Allium* and its effect on *Colletotrichum*. The same could perhaps apply to parasitism by *Peronospora destructor*, but environmentally-caused differences in metabolite levels have received little attention. The interaction between the physiological activities of leaves, shoots, and roots upon secondary metabolite production and translocation could have a marked effect upon disease establishment. In particular, the distribution of metabolites in clones of a host that have been grown in different communities, especially root rhizosphere and hyphosphere communities, has rarely been examined (but see Paxton, 1983). From the coevolutionary point of view, genotypic variation is less important than this community structure, its predictability, and its influence on metabolite production and accumulation in target organs. What is clear is that the study of the physiology of host–parasite relationships has not so far yielded hints as to the coevolution of the species-specific relationship.

Since the protoplasms of an haustorial obligate parasite and a host are never in contact, there must be a dynamic equilibrium between the components of the plasma membranes to control the differential permeabilities. The situation may be different for facultative parasites. The metabolism of component phospholipids and sterols of the plasma membranes could obviously be involved.

There are ample data to support the assumption that most of the Peronosporomycetidae have a dependence for intermediates of sterol metabolism (Warner *et al.*, 1982; Elliott, 1983). The steroid chemistry of the Oomycetes has not been fully investigated, but is clearly different from that of true fungi, as is shown by the actions of sterol-inhibiting fungicides (Fletcher, 1987) and polyene antibiotics.

There seems to be a connection between phytoalexin phenomena (which have been explored primarily at the level of the individual reactions between a parasite and its host), the sterol requirements of the parasitic species, and the secondary metabolites characteristic of the host order. The key could perhaps be an intense sterol-precursor requirement of the oomycete that can be satisfied by certain hosts which, for various reasons, have accumulations of a variety of secondary metabolites such as essential oils, saponins, and alkaloids. Secondary metabolite production could be stimulated by high carbohydrate levels or C:N ratios that would otherwise be excessive. The kind of photosynthetic pathway (C_4, as in sugarcane and sweetcorn), or high insolation (mediterranean plants, J. D. Ross, personal communication), or absence of a mycorrhizal carbon "sink" could all individually or collectively give rise to this situation. Sterol biosynthesis is linked through the terpenoids that relate to all these secondary metabolites. Paradoxically, the production of secondary metabolites may have been evolutionarily conserved because these deterred fungal parasites and predators. But the Oomycetes are not true fungi and as suggested below, the downy mildews may be of recent origin.

Investigations with *Phytophthora* (Hohl, 1983) have failed to reveal significant changes in sterol concentration or composition in relation to intraspecific variations in resistance or virulence. Stössel and Hohl (1981) found that high sterol ratios tend to correlate with susceptibility; this would be expected if host sterols were crucial to the evolution of the parasitic relationship. If this is so, the chemical diversity of the phytoalexins elicited in legumes (isoflavonoids) and solanums (terpenoids) may be explained. The phytoalexin phenomenon could be regarded as a by-product of parasite-induced diversion of host-steroid metabolism. Clarke (1983) has also suggested that phytoalexins are of secondary importance in hypersensitive reactions.

Discussions of differences in phytoalexin elicitation reflect abnormalities and deviations in the coevolutionary relationship. In natural populations, phytoalexin elicitation may confer a biological advantage *to the parasite* by limiting the extent of parasitism so that perennation of heterogeneous gene pools of both parasite and host is maintained. Only in artificial systems of clonal culture will this mechanism assume undue prominence.

Discussions of single-gene host resistance in different systems of host resistance and pathogen virulence (e.g., Keen and Yoshikawa, 1983) ignore the attraction and stimulation that enables both species to coexist. It is unlikely that studies concentrating on intraspecific differences will reveal underlying coevolutionary factors. There is an inverse relationship between the outlook and research momentum for plant pathology and the quest for an understanding of species-specific coevolution.

GEOGRAPHIC RADIATION

Geographic radiation is considered here with respect to the probabilities for coevolutionary and cophylogenetic radiation from geological evidence, and modern records of disease spread. Two important points of caution must be noted. First, for a few Peronosporomycetidae the current generic dispositions are questionable (e.g., Waterhouse, 1973). Secondly, a host may be atypical of the order or family, having evolved in a different region or at a different epoch from the higher taxon origin (e.g., *Thesium* (Santalaceae) a host of *Peronospora*); or a parasite may have become adapted to an "exceptional" host (e.g., *Peronospora destructor* on *Allium*).

The Cretaceous and Tertiary situation

In the Cretaceous (see Figure 6) there was a more or less continuous land mass

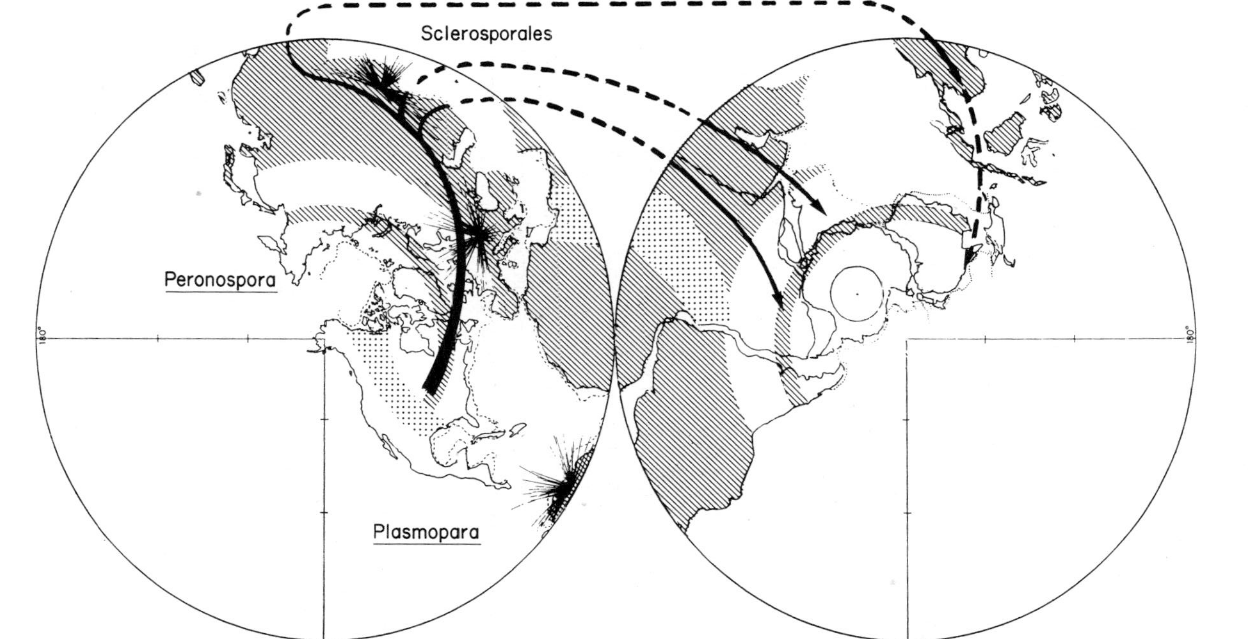

Figure 6. Cretaceous drift map, 100 ± 10 Myr. Present-day south pole, continental coastlines and shelves are outlined; the position of Cretaceous and Tertiary continental seaways are indicated by light stippling. Cretaceous angiosperm flora, mainly of woody dicotyledons, indicated by hatched bands (a) between the Tropics, and (b) for temperate circumpolar intercontinental continuities. Tertiary grazing migration routes (mammals and grasses) are indicated by arrows, solid (Miocene) and broken (Neogene). Outline map from Smith *et al.* (1973).

Cretaceous *Phytophthora* distribution would have approximated to the tropic dicotyledonous band, with various centres for speciation. *Albugo* distribution may have been comparable. Asters indicate centres of radiation during the Tertiary shown for *Plasmopara–Pseudoperonospora*, *Peronospora*, and the graminicolous downy mildews. For the graminicolous downy mildews, compare the grazing routes that became available from the tropics in the Miocene with the south circumpolar continental separations that provide barriers from the Palaeogene.

from the eastern North American continent in northern temperate latitudes to East Asia on the Tropic of Cancer (Laurentia–Laurasia, Angaraland–East Asia) (Tarling, 1980). The climate on the southern boundary was warm temperate to tropical. Northern South America was equatorial and was the last part of that continent to be separated from central West Africa as the South Atlantic Ocean opened up in the late Cretaceous. This ocean was probably connected by a seaway across central Africa to the Tethys Ocean to the north-east. The microplates of Central America drifted southwards after the Triassic, so that, with the Cretaceous American seaway from the Arctic to the Gulf of Mexico, there were north–south and east–west barriers to migration between the two American continents. The Central Atlantic and Tethys Oceans (including the Mediterranean Sea) were also a barrier to north–south migration, at least for larger animals (Coryndon and Savage, 1973), but perhaps not for angiosperms between West Africa and Spain.

Herbivorous hoofed mammals are considered to have coevolved with the pasture grasses (Clayton and Renvoize, 1986) in North America in the Tertiary or possibly late Cretaceous (Maastrichtian) as uplift eliminated the American seaway. Thus, at the start of the Tertiary, the African plate (West Africa with East Africa) was an isolated island, as was India. It was only in the Miocene that the land bridges or island chains enabled immigrant grazing mammals to colonize Africa from Europe via south Asia and west India (Coryndon and Savage, 1973). Emigration was not prevalent, except perhaps from east Africa to India. Australasia was isolated from Antarctica (Gondwanaland), drifting northwards with a temperate-adapted biota towards south-east Asia, so that it was only from the Oligocene that land bridges permitted tropical organisms a southward migration, which continued through the Quaternary.

For the angiosperm hosts of the Peronosporomycetidae (Figures 5 and 6) this meant that there was a more or less continuous belt of an essentially tropical dicotyledonous flora from South America through western Africa and southern Europe to east Asia. The pasture grasses must have radiated from the tropical/warm temperate climate of what is now temperate North America in the early Tertiary, moving southwards and eastwards as the climate cooled and habitats were opened up with the burgeoning of the grazing mammals. Since the graminicolous downy mildews are primarily associated with the highly evolved tropical grazing grasses of the Paniceae, Andropogoneae, the earlier origins of the bamboos, rices, and Southern Hemisphere grasses can be ignored here (see Clayton and Renvoize, 1986). Comparison of the present host ranges of the downy mildews and the relative antiquities of their hosts (Figure 7), together with what is known of host ranges for *Pythium* and *Phytophthora*, present some problems. The affinity of *Phytophthora* for woody dicotyledonous hosts and the distinct, later origin of the pasture grasses makes it possible to suggest that, contrary to accepted phylogenies (Barr, 1983; Shaw, 1981; Skalíky, 1966), *Pythium* may have evolved later than *Phytophthora*. This receives some support from mitochondrial genome analyses (Belkhiri, 1987) which indicate that *Phytophthora* lacks the inverted repeat characteristic of *Pythium* (McNabb *et al.*, 1987). While it is realistic to propose a late Cretaceous origin for *Phytophthora*, that of *Pythium* is more likely to have been with the pasture grasses in the Tertiary.

Albugo, on hosts of early origin (Amaranthaceae in Asia; Convolvulaceae in South America), may well be ancient, with a late Cretaceous origin in common with *Phytophthora*. *Peronospora* and *Albugo* share many of the host angiosperm orders. However, *Peronospora* is predominantly Northern Hemisphere. These two genera are also common on Brassicaceae and other late-evolving angiosperm orders, of which many members are herbaceous and found in conditions of high insolation (montane or mediterranean). An appreciable part of the modern mediterranean flora is thought to have been derived from North Africa at different times, through Spain, Asia Minor, or the Italian peninsula depending upon proximities in different epochs. This suggests a mid- to late-Tertiary origin for *Peronospora*.

The most plausible origin of *Plasmopara* (note hosts in Asteraceae, Vitaceae and Caprifoliaceae), and *Pseudoperonospora* (note hosts in Urticaceae and Cucurbitaceae) is from tropical South America in the mid- to late Palaeogene. Although there could have been early radiation through West Africa to Eurasia before the South Atlantic expanded, host origins do not indicate that this occurred. But, as the two American continents drew together (*c.* 50 million years ago, radiation through Central America was possible. It may be speculated that, unlike the South American flora, that of North America would have been particularly vulnerable to parasites evolving from *Phytophthora* because it would have had little opportunity to develop genes for resistance to this genus.

In contrast to the dicotyledonous downy mildews, the graminicolous downy mildews are distinct, cophylogenetically linked, and could have evolved when grazing became a selective force for the tropical grasses of south Asia. A "pro-*Pythium*" ancestor might be presumed, but there are indications that a saprolegniomycetidous origin is also likely. Biochemical evidence (Belkhiri, 1987; G. R. Klassen, personal communication) does not support a link between *Verrucalvus* or *Pachymetra* with either *Phytophthora* or *Pythium* and mitochondrial DNA restriction patterns in *Verrucalvus* are quite close to those of the Saprolegniaceae. In this context, the grassland affinities of *Aplanopsis*, and particularly *Leptolegnia caudata*, should be noted.

The Himalayan uplift, which drained the Tethys Ocean remnants and resulted in subsequent colder montane conditions, would have forced the

Figure 7. Distribution of downy mildews with respect to host orders, with approximate geological origins of families parasitized, based upon the compilation by Muller (1981) of pollen data. The pecked horizontal lines correspond to the supra-orders in Figures 4 and 5. Three levels are given in the Neogene corresponding to Lower Miocene (>15 Myr), Upper Miocene (<11 Myr), and Pliocene (<5 Myr) respectively.

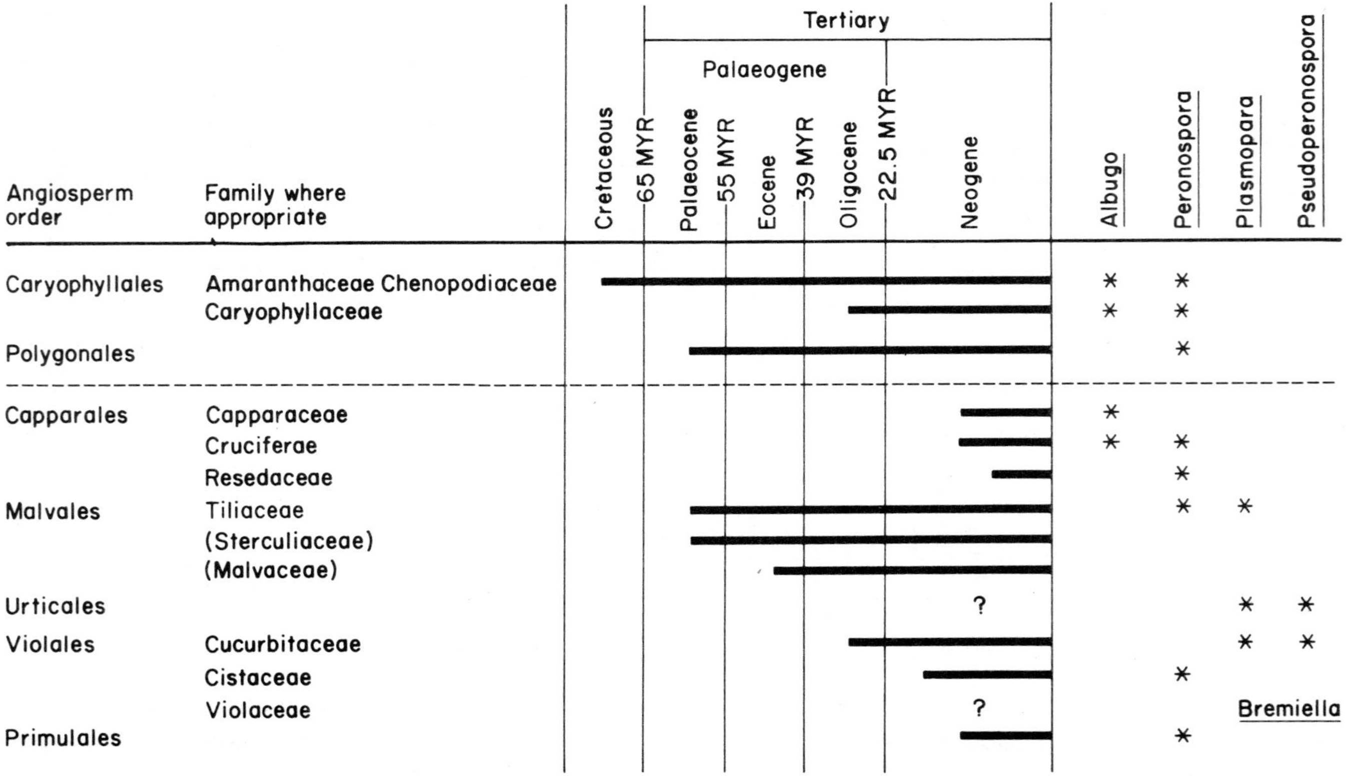

Tertiary: Palaeogene (Cretaceous — 65 MYR — Palaeocene — 55 MYR — Eocene — 39 MYR — Oligocene — 22.5 MYR) — Neogene

Angiosperm order	Family where appropriate	Neogene	Albugo	Peronospora	Plasmopara	Pseudoperonospora
Caryophyllales	Amaranthaceae Chenopodiaceae		*	*		
	Caryophyllaceae		*	*		
Polygonales				*		
Capparales	Capparaceae		*			
	Cruciferae		*	*		
	Resedaceae			*		
Malvales	Tiliaceae			*	*	
	(Sterculiaceae)					
	(Malvaceae)					
Urticales		?			*	*
Violales	Cucurbitaceae				*	*
	Cistaceae			*		
	Violaceae	?			Bremiella	
Primulales				*		

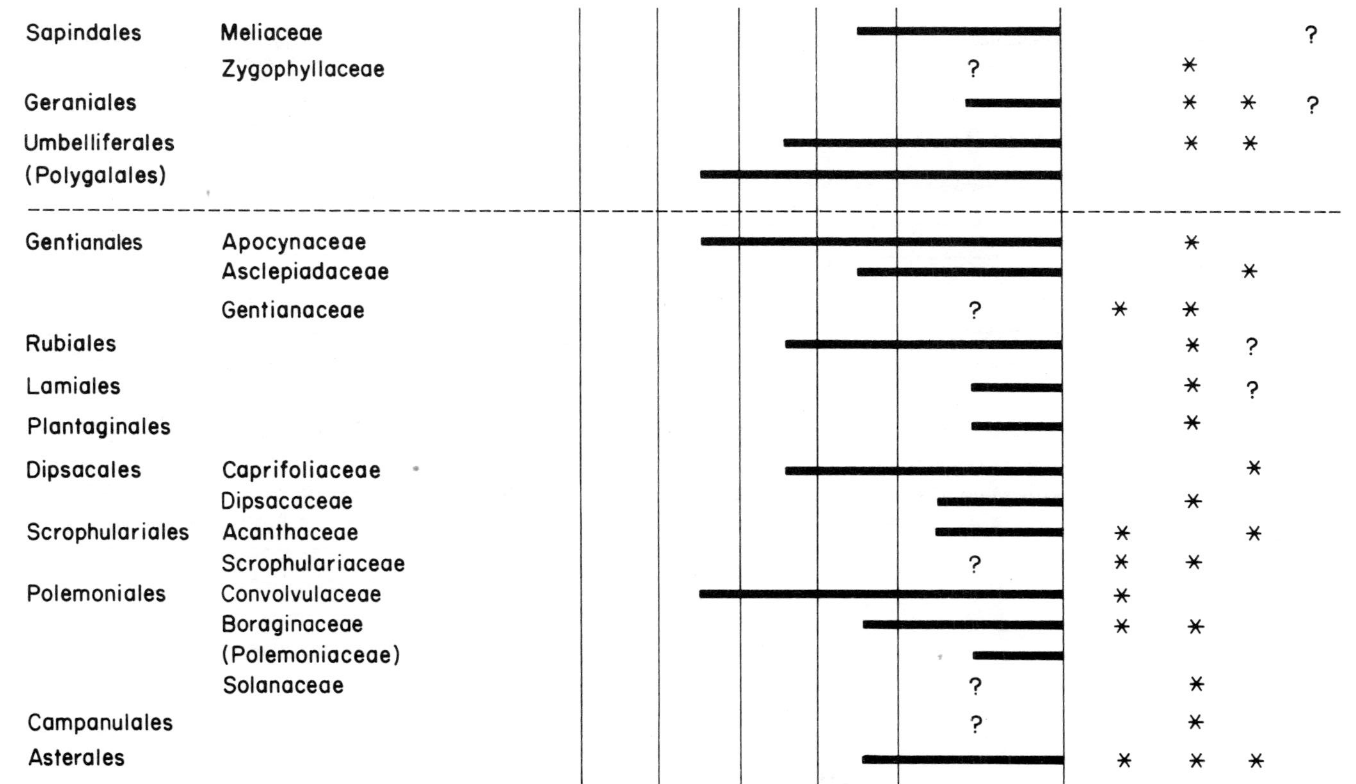

Sapindales
Meliaceae
Zygophyllaceae
Geraniales
Umbelliferales
(Polygalales)
Gentianales
Apocynaceae
Asclepiadaceae
Gentianaceae
Rubiales
Lamiales
Plantaginales
Dipsacales
Caprifoliaceae
Dipsacaceae
Scrophulariales
Acanthaceae
Scrophulariaceae
Polemoniales
Convolvulaceae
Boraginaceae
(Polemoniaceae)
Solanaceae
Campanulales
Asterales

Angiosperm order	Family where appropriate	Cretaceous	Palaeocene (Tertiary, Palaeogene; 65–55 MYR)	Eocene (Tertiary, Palaeogene; 55–39 MYR)	Oligocene (Tertiary, Palaeogene; 39–22.5 MYR)	Neogene (Tertiary)	Albugo	Peronospora	Plasmopara	Pseudoperonospora
Rosales	Rosaceae				—	—		*		
	Crassulaceae					?	*			
	Saxifragaceae					?		*	*	
Fabales	Fabaceae[1]					—	*	*		
Myrtales	Onagraceae	—	—	—	—	—		*	*	
Santalales	Santalaceae		—	—	—	—		*		
Rhamnales	Vitaceae				—	—			*	
Euphorbiales	Euphorbiaceae	—	—	—	—	—		*		
Piperales						?	*			
Ranunculales						—		*	*	
Papaverales						?		*		
Poales			—	—	—	—			??[2]	
Liliales				—	—	—		*[3]		

1. Many Fabales are of much more ancient origin.
2. *Plasmopara olpismene* and *P. penniseti* may be wrongly assigned to this genus. *Basidiophora butleri* has been returned to *Sclerospora* (Dick *et al.*, 1984).
3. *Peronospora destructor* only.

early graminicolous downy mildews southwards into Africa, India, and south east Asia. This pattern of change could also explain the new-found occurrence of the north-eastern Australian genera *Verrucalvus* on the introduced Kikuyu grass *Pennisetum clandestinum* (Dick *et al.*, 1984) and *Pachymetra* on sugarcane (Dick *et al.*, 1988). The absence of any associations with southern temperate grasses makes it unlikely that there was early evolution of these graminicolous fungi from Gondwanaland, with isolated three-pronged migration and evolution in Africa, India, and Australasia (but not southern South America). Only if the Sclerosporales are considered to be of late Neogene or Quaternary origin could a centre of radiation from Africa (Williams, 1984) or India be supported. Such an origin is also unlikely on morphological evidence from the fungi or biogeography of the grasses.

The evolutionary pattern of the Peronosporales, which shows even more adaptive radiation (*Basidiophora*, *Bremia*) in the Asteraceae, is not primarily cophylogenetic. It is probably relatively recent, mostly developing during the late Tertiary, less than 10 million years ago. It could be inferred that, within the Peronosporomycetidae, morphology, particularly in relation to zoospore production, is an unreliable indicator of phylogenetic age.

The absence of direct phylogenetic sequences between the evolution of *Plasmopara*, *Peronospora*, or the Sclerosporales suggests that further revision of family boundaries in the Peronosporales might be appropriate.

The history of disease spread

In historic time, man has been a major influence on the distribution of downy mildews. Since agriculture, exploration, and plant introductions pre-date scientific information on pathogens, the localities from which mildews were first described are not necessarily indicative of where they originated. The information on epidemiological spread of *Phytophthora infestans* on potato in Europe in the nineteenth century (Gregory, 1983), and in this century of *Peronospora tabacina* on tobacco in North America and Europe and *Sclerospora graminicola* on agricultural grasses in south-east Asia (Welzien, 1981) is more instructive. Under agriculture the spread of these parasites across continents occurred within a few decades. The introduction of the susceptible New World *Zea* into Europe and Asia provides a good example of how quickly the downy mildews can adapt to a newly presented host. Ball and Pike (1984) showed that while fairly stable host–pathogen relationships have developed in Pearl Millet (*Pennisetum americanum*) since its introduction into India about 3000 years ago, the intercontinental variation in pathogenicity in *Sclerospora graminicola* suggests that appreciable intraspecific co-evolution can occur in equivalent timescales. Speciation in *Phytophthora*,

faunas. In "Organisms and Continents Through Time" (N. F. Hughes, ed.), pp. 121–135. Palaeontological Association, London.

Crute, I. R. (1981). The host specificity of peronosporaceous fungi and the genetics of the relationship between host and parasite. In "The Downy Mildews" (D. M. Spencer, ed.), pp. 237–253. Academic Press, London.

Deacon, J. W. (1976). Studies on *Pythium oligandrum* an aggressive parasite of other fungi. *Trans. Br. mycol. Soc.* **66,** 383–391.

Dick, M. W. (1969). Morphology and taxonomy of the Oomycetes, with special reference to Saprolegniaceae, Leptomitaceae and Pythiaceae. I. Sexual reproduction. *New Phytol.* **68,** 751–775.

Dick, M. W. (1976). The ecology of aquatic phycomycetes. In "Recent Advances in Aquatic Mycology" (E. B. G. Jones, ed.), pp. 513–542. Elek Press, London.

Dick, M. W. (1987). Sexual reproduction: nuclear cycles and life-histories with particular reference to lower eukaryotes. *Biol. J. Linn. Soc.* **30,** 181–192.

Dick, M. W. (1988). Oomycota. In "Handbook of Protoctista" (L. Margulis, J. O. Corliss, M. Melkonian, and D. Chapman, eds). Jones and Bartlett, Boston, in press.

Dick, M. W. and Hawksworth, D. L. (1985). A synopsis of the biology of the Ascomycotina. *Bot. J. Linn. Soc.* **19,** 175–179.

Dick, M. W., Wong, P. T. W., and Clark, G. (1984). The identity of the oomycete causing "Kikuyu Yellows", with a reclassification of the downy mildews. *Bot. J. Linn. Soc.* **89,** 171–197.

Dick, M. W., Croft, R. J., Magarey, R. C., de Cock, A. W. A. M., and Clark, G. (1988). *Pachymetra*, a new genus of the Verrucalvaceae (Oomycetes). *Bot. J. Linn. Soc.*, in press.

Dilcher, D. L. (1965). Epiphyllous fungi from Eocene deposits in western Tennessee, USA. *Palaeontographica* **116,** 1–54.

Elad, Y. and Chet, I. (1987). Possible role of competition for nutrients in biocontrol of *Pythium* damping-off by bacteria. *Phytopathology* **77,** 190–195.

Elliott, C. G. (1983). Physiology of sexual reproducion in *Phytophthora*. In "*Phytophthora*, Its Biology, Taxonomy, Ecology and Pathology" (D. C. Erwin, S. Bartnicki-Garcia and P. H. Tsao, eds), pp. 71–80. American Phytopathological Society, St Paul, Minn.

Fletcher, R. A. (1987). Plant growth regulating properties of sterol-inhibiting fungicides. In "Hormonal Regulation of Plant Growth and Development" (S. S. Purchit, ed.), pp. 103–114. Martinus Nijhoff, Dordrecht.

Galston, A. W. (1983). Polyamines as modulators of plant development. *BioScience* **33,** 382–388.

Gerdemann, J. W. (1968). Vesicular-arbuscular mycorrhiza and plant growth. *A. Rev. Phytopath.* **6,** 397–418.

Gleason, F. (1976). The physiology of the lower freshwater fungi. In "Recent Advances in Aquatic Mycology" (E. B. G. Jones, ed), pp. 543–572. Elek Press, London.

Gregory, P. H. (1983). Some major epidemics caused by *Phytophthora*. In "*Phytophthora*, Its Biology, Taxonomy, Ecology and Pathology" (D. C. Erwin, S. Bartnicki-Garcia, and P. H. Tsao, eds), pp. 271–278. American Phytopathological Society, St Paul, Minn.

Hardman, J. M. (1985). "The ecology of *Pythium*: propagule survival and retrieval in natural and artificial systems". PhD thesis, University of Reading.

Hohl, H. R. (1983). Nutrition of *Phytophthora*. In "*Phytophthora*, Its Biology, Taxonomy, Ecology and Pathology" (D. C. Erwin, S. Bartnicki-Garcia, and P. H. Tsao, eds), pp. 41–54. American Phytopathological Society, St Paul, Minn.

Holwill, M. E. J. (1982). Dynamics of eukaryotic flagellar movement. In "Prokaryotic and Eukaryotic Flagella" (W. B. Amos and J. G. Duckett, eds), pp. 289–312. Cambridge University Press, Cambridge.

Ingram, D. S. (1981). Physiology and biochemistry of host–parasite interaction. In "The Downy Mildews" (D. M. Spencer, ed.), pp. 143–163. Academic Press, London.

Karling, J. S. (1981). "Predominantly Holocarpic and Eucarpic Simple Biflagellate Phycomycetes". Cramer, Vaduz.

Keen, N. T. and Yoshikawa, M. (1983). Physiology of disease and the nature of resistance to *Phytophthora*. In "*Phytophthora*, Its Biology, Taxonomy, Ecology and Pathology" (D. C. Erwin, S. Bartnicki-Garcia, and P. H. Tsao, eds), pp. 279–287. American Phytopathological Society, St Paul, Minn.

Kevan, P. G., Chaloner, W. G., and Savile, D. B. O. (1975). Interrelationships of early terrestrial arthropods and plants. *Palaeontology* **18,** 391–417.

McNabb, S. A., Boyd, D. A., Belkhiri, A., Dick, M. W., and Klassen, G. R. (1987). An inverted repeat comprises more than three-quarters of the mitochondrial genome in two species of *Pythium*. *Current Genetics* **12,** 205–208.

Martin, R. W. and Miller, C. E. (1985). Host range and taxonomy of *Olpidiopsis varians*. *Mycotaxon* **24,** 411–418.

Moore, P. D. (1982). Evolution of photosynthetic pathways in flowering plants. *Nature, Lond.* **295,** 647–648.

Morrall, S. and Greenwood, A. D. (1982). Ultrastructure of nucleomorph division in species of Cryptophyceae and its evolutionary implications. *J. Cell Sci.* **54,** 311–328.

Müller, D. G. and Falk, H. (1973). Flagellar structure of the gametes of *Ectocarpus siliculosus* (Phaeophyta) as revealed by negative staining. *Arch. Mikrobiol.* **91,** 313–322.

Muller, J. (1981). Fossil pollen records of extant angiosperms. *Bot. Rev.* **47,** 1–142.

Newell, W. Y., Cefalu, R. and Fell, J. W. (1977). *Myzocytium*, *Haptoglossa*, and *Gonimochaete* (Fungi) in littoral marine nematodes. *Bull. mar. Sci.* **27,** 177–207.

Palti, J. and Kenneth, R. (1981). The distribution of downy mildew genera over the families and genera of higher plants. In "The Downy Mildews" (D. M. Spencer, ed.), pp. 45–56. Academic Press, London.

Paxton, J. D. (1983). *Phytophthora* root rot and stem rot of soybean: a case study. In "Biochemical Plant Pathology" (J. A. Callow, ed.), pp. 19–30. Wiley, New York.

Pirozynski, K. A. (1976). Fossil fungi. *A. Rev. Phytopath.* **14,** 237–246.

Pirozynski, K. A. and Malloch, D. W. (1975). The origin of land plants: a matter of mycotrophism. *BioSystems* **6,** 153–164.

Powell, M. J., Lehnen, L. P., and Bortnick, R. M. (1985). Microbody-like organelles as taxonomic markers among Oomycetes. *BioSystems* **18,** 321–334.

Sackston, W. E. (1981). Downy mildew of sunflower. In "The Downy Mildews" (D. M. Spencer, ed.), pp. 545–575. Academic Press, London.

Sansome, E. and Sansome, F. W. (1974). Cytology and life-history of *Peronospora parasitica* on *Capsella bursa-pastoris* and of *Albugo candida* on *C. bursa-pastoris* and on *Lunularia annua*. *Trans. Br. mycol. Soc.* **62,** 323–332.

Shaw, C. G. (1981). Taxonomy and evolution. In "The Downy Mildews" (D. M. Spencer, ed.), pp. 17–29. Academic Press, London.

Skalíky, C. V. (1966). Taxonomie der Gattungen der Familie Peronosporaceae. *Preslia* (*Praha*) **38,** 117–129.

Skidmore, D. I. and Ingram, D. S. (1985). Conidial morphology and the specialization of *Bremia lactucae* Regel (Peronosporaceae) on hosts in the family Compositae. *Bot. J. Linn. Soc.* **91,** 503–522.

Slifkin, M. K. (1961). Parasitism of *Olpidiopsis incrassata* on members of the Saprolegniaceae. I. Host range and effects of light, temperature, and stage of host on infectivity. *Mycologia* **53,** 183–193.

Smith, A. G., Briden, J. C., and Drewry, G. E. (1973). Phanerozoic world maps. In "Organisms and Continents Through Time" (N. F. Hughes, ed.), pp. 1–42. Palaeontological Association, London.

Sparrow, F. K. (1960). "Aquatic Phycomycetes", 2nd Edn. University of Michigan Press, Ann Arbor.

Stanghellini, M. E. (1974). Spore germination, growth and survival of *Pythium* in soil. *Proc. Am. Phytopath. Soc.* **1,** 211–214.

Stebbins, G. L. (1974). "Flowering Plants—Evolution above the Species Level". Arnold, London.

Stössel, P. and Hohl, H. R. (1981). Sterols in *Phytophthora infestans* and their role in the parasitic interactions with *Solanum tuberosum*. *Ber. schweiz. bot. Ges.* **90,** 118–128.

Tarling, D. H. (1980). "Continental Drift and Biological Evolution" (Carolina Biology Readers, 113). Biological Supply Company, Burlington, Carolina.

Tarr, S. A. J. (1972). "The Principles of Plant Pathology", p. 251. Macmillan, London.

Vaziri-Tehrani, B. and Dick, M. W. (1980). Amino acid composition of oomycete cell walls. *Trans. Br. mycol. Soc.* **74,** 225–230.

Warner, S. A., Eierman, D. F., Savocool, G. W., and Domnas, A. J. (1982). Cycloartenol-derived sterol biosynthesis in the Peronosporales. *Proc. natn. Acad. Sci. U.S.A.* **79,** 3769–3772.

Waterhouse, G. M. (1973). Peronosporales. In "The Fungi" (G. C. Ainsworth, F. K. Sparrow, and A. S. Sussman, eds), vol. 4B, pp. 165–183. Academic Press, New York.

Welzien, H. C. (1981). Geographical distribution of downy mildews. In "The Downy Mildews" (D. M. Spencer, ed.), pp. 31–43. Academic Press, London.

Whitfield, P. R. and Bottomley, W. (1983). Organization and structure of chloroplast genes. *A. Rev. Pl. Physiol.* **34,** 279–310.

Williams, R. J. (1984). Downy mildews of tropical cereals. In "Advances in Plant Pathology (D. S. Ingram and P. H. Williams, eds.), vol. 2, pp. 1–103. Academic Press, London.

Wolf, F. A. (1966). Fungus spores in East African lake sediments. *Bull. Torrey bot. Club* **93,** 104–113.

Wolf, F. A. (1968). Fungus spores in Lake Singletary sediment. *J. Elisha Mitchell Sci. Soc.* **84,** 227–232.

Yarwood, C. E. (1977). *Pseudoperonospora cubensis* in rust-infected bean. *Phytopathology* **67,** 1021–1022.

Zentmyer, G. A. (1980). "*Phytophthora cinnamomi* and the Diseases it Causes". American Phytopathological Society, St Paul, Minn.

3 | Coevolution of Flowering Plants with Pathogenic Fungi

T. HIJWEGEN

Department of Phytopathology, Agricultural University, Wageningen, The Netherlands

Abstract

Coevolution of flowering plants with pathogenic fungi is discussed with emphasis on coevolution in time rather than in space from the phylogenetic point of view. Special attention is devoted to the problem of coevolution and "jumps" to disparate hosts. The importance of "jumps" in crop protection is emphasized and the possibilities of making use of host–parasite data in the biological control of weeds is indicated. The role of fungi as taxonomists is described. Some examples of coevolution and "jumps" are elaborated and the processes involved are discussed.

INTRODUCTION

The coevolution of biotrophic fungi and their host plants is an intriguing subject, especially when used to deduce relationships and construct phylogenies elaborating the system in two directions, one for the pathogens and one for the host plants, when the two would be congruent.

Although various authors have made important contributions to this subject, the art of utilizing host–parasite data as an aid to taxonomy and evolution is primarily due to Savile, who published many papers on this subject, culminating in a comprehensive treatise (Savile, 1979).

"Trying to trace the evolution and geographic history of a group of obligate parasites, without consideration of their hosts, must generally be a somewhat sterile pursuit, for the rusts can only follow their hosts wholly or in part. But occasionally a combined study of the two coevolving groups may be doubly illuminating: first, we have the maximum pool of data to draw upon, so that indications from the parasites may strengthen others derived from the hosts and, second, the relatively simple morphology of the parasites often allows us to arrange them in evolutionary sequence much more easily than we can do for the hosts." (Savile, 1975, p. 354).

COEVOLUTION OF FUNGI
WITH PLANTS AND ANIMALS
ISBN 0-12-557365-0

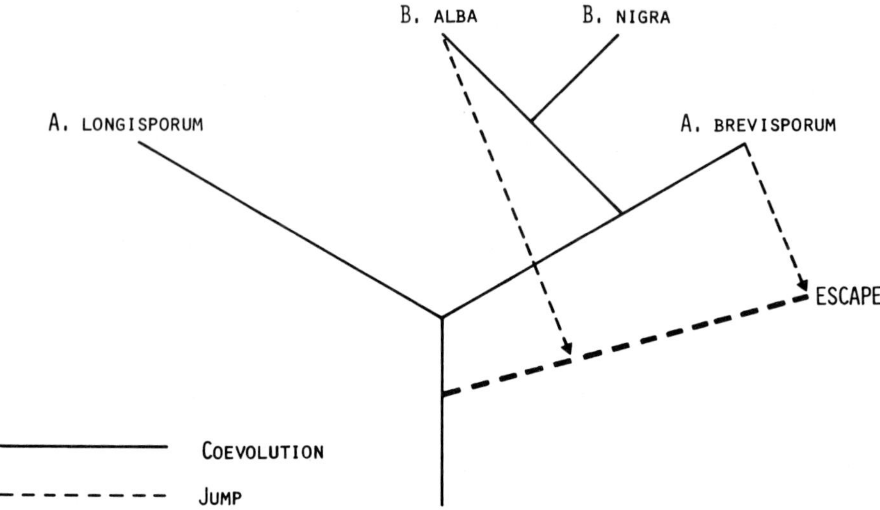

Figure 1. Diagrammatic representation of coevolution and jumps. Four hypothetical fungi evolved on their respective host plants. Genus "A" with two species differing in spore size and genus "B" with two species with differently coloured spores. All four may jump to an escaped population of plants, which is depicted for two species.

Coevolution is understood here in the sense of Savile, that is, evolution in conjunction with the evolving host (Parlevliet, 1986); a fungus or a group of fungi thriving on a host plant or a group of host plants that, starting at a certain point in evolutionary time, have been evolving with each other ever since. In other words, the coevolution of plants and biotrophic fungi is equivalent to the evolution of a pathosystem (a combination of plant and pathogen) giving rise to a branched system (Figure 1).

In pathosystems, the benefit is unidirectional: the pathogen exploits the host. There is no known mutual benefit, as in the legume–*Rhizobium* or flower–butterfly associations (Ehrlich and Raven, 1964), so the mutualistic definition of coevolution is not pertinent to these cases.

"Jumps", which can be defined as transfers from one host species to another, can be either physical—without any genetic change (Hijwegen, 1979)—or with genetic change (e.g., Savile, 1969). In this contribution, attention is focused on jumps without genetic change, but it is usually difficult to discriminate between the two cases, especially when a jump occurred in the distant past. I am increasingly convinced that coevolution is relatively rare in pathosystems and that most lineages described are, at best, a mixture of coevolutionary events and jumps to closely related taxa. This same conviction is expressed in reference to rusts of Triticeae by Baum and Savile (1985). However, although in many instances complete coevolution cannot be accepted, it must be assumed, that, in any case, a particular set of genes for compatibility is involved that is not present in other groups of plants.

These genes for compatibility, whether unmasked, masked, or altered, are the "red thread" in the story. Sequencing of genes for compatibility might be as important to the plant pathologist as to the taxonomist interested in phylogenetic problems. Unfortunately, no such gene has currently been identified.

Since numerous papers have already appeared on the art of making use of host–parasite data in plant taxonomy, this paper concentrates on the problem of coevolution and jumps and the dynamics of the process.

ORIGINS: ADAPTATION AND SPECIALIZATION

Erysiphe martii is a powdery mildew reported to occur in some genera of Leguminosae (Blumer, 1967); this fungus has both specialized and unspecialized races. Unspecialized races may be transferred from one host species to another of quite a different genus with concomitant adaptation, usually by single or double saltation, suggesting a rearrangement of genetic material. On a particular host (e.g., *Trifolium incarnatum*), *E. martii* will thrive and will sporulate 4–6 days after germination but, after transfer to *Lathyrus odoratus*, hyphae are formed that are thinner than originally and without conidiophores. From (one of) these thinner hyphae a thicker hypha may originate bearing conidiophores (single saltation). If no conidiophores arise from these somewhat thicker hyphae, still thicker hyphae may be formed to bear conidiophores (double saltation). Conidia may arise 7–10 days after germination. When kept on the same host, the adapted race will sporulate normally after 4–6 days. On transfer to the original host, the whole process is reversed, complete with a lag phase (Hijwegen, unpublished).

In July 1984, an isolate of *E. martii* was obtained from *Trifolium incarnatum* that after adaptation, also could infect, *Lathyrus odoratus* and *Lupinus luteus*. In July 1985, after 25 transfers in the greenhouse on *T. incarnatum*, the strain was no longer able to infect *L. odoratus* but remained infectious to *L. luteus*, although without sporulation. The fungus had undergone specialization. It can be envisaged that in this way a lineage of related fungi can evolve on a related group of host plants.

Specialization with a loss of adaptability is evolutionarily risky, especially in the case of annual plants such as *T. incarnatum*. Therefore, it is not surprising that despecialization also occurs: in autumn, when the powdery mildew is abundant, genetic material from different populations may come together and perithecia may form; in the spring the process of specialization may start again. Neger (1901, 1902) noted that in powdery mildews the anamorph,

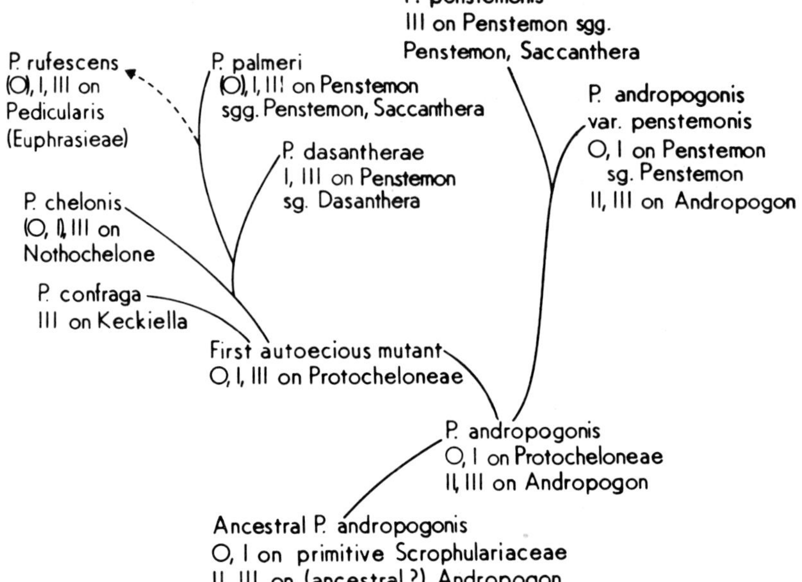

Figure 2. Diagram showing the probable relationships (connecting lines) and relative chronology (vertical positions) of rusts attacking Cheloneae. Modified from Savile (1969) with permission. 0 = spermagonia, I = aecia, II = uredinia, III = telia.

according to Tranzschel's law. Its host relationship suggests that it arose much later than the other group (Savile, 1969).

Evolution of this group of rusts has occurred in temperate North America. However, sometimes geographical migrations complicate the reconstruction of the history of the evolution of pathosystems. An elegant example of such a reconstruction is that of the evolution of the Saxifragaceae s.s. as mirrored by microcyclic *Puccinia* species (Savile, 1975). Savile concluded that, judging from the rust data, the family originated in eastern or south-eastern Asia with an ancestral plant that may have looked much like *Astilbe*. The most ancient extant pathosystem is that of *P. tottoriensis* and *Mitella* in Japan. The first radiations harbouring the most primitive rusts of this lineage were all Asiatic and gave rise to the ancestors of *Tiarella* (with *P. tiarellicola*) and *Mitella* (with *P. tottoriensis* and *P. asiatica*) in eastern Asia, and probably to *Chrysosplenium* in the Himalayan region. Another branch apparently crossed the Bering Bridge to give rise to the beginnings of various North American genera with *P. heucherae* and its varieties on *Heuchera* and related genera. One product of this cordilleran radiation must have been a "*Protosaxifraga*" that recrossed the Bridge and radiated in or near Japan, where the most primitive *Saxifraga* species are found. (Subsequent evolution of these rusts is on *Saxifraga*, with "jumps" to other genera.) This complex sent branches to the Himalayan region and from there at least one radiant reached Europe. A slightly later radiant from Japan entered Beringia, where radiation gave slightly more advanced saxifrages harbouring *P. austro-beringiana*. Another migrant from Japan later radiated in the Himalayas producing *P. saxifragae* var. *saxifragae* on *Melanocentrae*. From this complex, a branch reached Europe and a later branch reached North America probably via the Siberian coast to produce subsect. *Nivali-Virginiensis*. In the meantime, *Saxifraga* in the Himalayas was moving into more arid habitats giving rise to sect. *Kabschia*; radiants from the Himalayan *Kabschia* section finally gave rise to the most modern sections found in the Alps and Beringia.

Similar evolutionary histories may be reconstructed for other rust genera (Figure 3). Species of the rust genus *Uropyxis* (Baxter, 1959; Hennen and Cummins, 1973) are reported from Africa and the Americas on two closely related tribes in the Leguminosae-Papilionioideae, Amorpheae, and Aeschynomeneae, and on the genus *Diphysa* (of somewhat doubtful affinities, now included in Robinieae but perhaps more closely related to Aeschynomeneae; Polhill and Sousa, 1981; Hijwegen, 1982*). This lineage was most probably established by a full-cycled autoecious *Diorchidium*-like rust on the ancestor of Aeschynomeneae and Amorpheae somewhere between West Africa and Mexico, after the separation of South America but prior to the separation of North America 49 million years ago (Raven and Polhill, 1981). The most primitive extant representatives, though now probably microcyclic, are *Uropyxis steudneri* on *Ormocarpum bibracteatum* in Ethiopia and its var. *rhodesica* on *O. trichocarpum* in Zambia (both with thick-walled diorchidioid teliospores). The main evolution was probably predominant in the North American continent. The species with the closest resemblance to *U. steudneri* are the also microcyclic *U. diphysae* on *Diphysa* species in Central America (with two types of telia) and *U. holwayi* on *D. floribunda* and *Eysenhardtia* species in Central America (with considerable variability in size and colour of teliospores and in colour of telia). The fungus probably "jumped" to *Eysenhardtia* species, since this genus is also compatible with another rust and could be termed a "collector" (Gäumann, 1959). Subsequent evolution has evidently been predominantly on Amorpheae. A somewhat aberrant species is *U. amorphae* on *Amorpha* species and *Parryella filifolia* in the USA; this has a very thick hygroscopic outer layer in the teliospore wall.

Starting with *U. daleae*, species appear to show a loss of morphological variability; this species probably evolved on either *Dalea* or *Eysenhardtia* in Central America and "jumped" to Aeschynomeneae in W. Africa when the

---

* *Diphysa* was transferred to Aeschynomeneae subtr. Ormocarpine by Lavin (in "Advances in Legume Systematics, Part 3" (C. H. Stirton, ed), pp. 31–64, 1987. Royal Botanic Gardens, Kew.

Anikster, Y. and Wahl, I. (1979). Coevolution of the rust fungi on Gramineae and Liliaceae and their hosts. *A. Rev. Phytopath.* **17,** 367–403.

Arthur, J. C. (1934). "Manual of the Rusts in United States and Canada". Purdue Research Foundation, Lafayette.

Baum, B. R. and Savile, D. B. O. (1985). Rusts (Uredinales) of Triticeae: evolution and extent of coevolution, a cladistic analysis. *Bot. J. Linn. Soc.* **91,** 367–394.

Baxter, J. W. (1959). A monograph of the genus *Uropyxis. Mycologia* **51,** 210–226.

Blumer, S. (1967). "Echte Mehltaupilze (Erysiphaceae)". Fischer, Jena.

Boerema, G. H. and Verhoeven, A. A. (1972). Check-list for scientific names of common parasitic fungi. Series 1a: Fungi on trees and shrubs. *Neth. J. Pl. Path.* **78** (Suppl. 1), 1–63.

Boerema, G. H., Loerakker, W. M. and Kesteren, H. A. van (1983). Valse meeldauw bij *Lisianthus russellianus. Jaarb. PlantenziektenKde Dienst* **1983,** 16–18.

Boulter, D. (1976). The evolution of plant proteins with special reference to higher plant cytochromes *c*. In "Commentaries in Plant Science" (H. Smith, ed.), pp. 77–89. Pergamon Press, Oxford.

Bruzzese, E. and Hasan, S. (1986*a*). The collection and selection in Europe of isolates of *Phragmidium violaceum* (Uredinales) pathogenic to species of European black berry naturalised in Australia. *Ann. appl. Biol.* **108,** 527–533.

Bruzzese, E. and Hasan, S. (1986*b*). Host specificity of the rust *Phragmidium violaceum*, a potential biological control agent of European blackberry. *Ann. appl. Biol.* **108,** 585–596.

Cummins, G. B. (1971). "The Rust Fungi of Cereals, Grasses and Bamboos". Springer-Verlag, Berlin.

Cummins, G. B. (1978). "Rust Fungi on Legumes and Composites in North America". University of Arizona Press, Tucson.

Dahlgren, R. M. T. (1980). A revised system of classification of the angiosperms. *Bot. J. Linn. Soc.* **80,** 91–124.

Dinoor, A. and Wahl, I. (1963). Reaction of non-cultivated oats from Israel to Canadian races of crown rust and stem rust. *Can. J. Pl.Sci.* **43,** 468–470.

Ehrlich, P. R. and Raven, P. H. (1964). Butterflies and plants: a study in coevolution. *Evolution* **18,** 586–608.

Eshed, N. and Dinoor, A. (1980). Genetics of pathogenicity in *Puccinia coronata*: Pathogenic specialization at the host genus level. *Phytopathology* **70,** 1042–1046.

Eshed, N. and Wahl. I. (1970). Host ranges and interrelations of *Erysiphe graminis hordei, E. graminis tritici* and *E. graminis avenae. Phytopathology* **60,** 628–634.

Gäumann, E. (1959). Die Rostpilze Mitteleuropas. *Beitr. KryptogFl. Schweiz* **11,** 1–1407.

Hasan, S. (1972). Specificity and host specialization of *Puccinia chondrillina. Ann. appl. Biol.* **72,** 257–263.

Hennen, J. F. and Cummins, G. B. (1973). New taxa of Mexican fungi. *Rep. Tottori mycol. Inst.* **10,** 169–182.

Hijwegen, T. (1979). Fungi as plant taxonomists. *Symb. bot. upsal.* **22**(4), 146–165.

Hijwegen, T. (1981). Fungi as plant taxonomists. II. Affinities of the Rosiflorae. *Acta bot. neerl.* **30,** 479–491.

Hijwegen, T. (1982). The rust genus *Uropyxis* and the position of *Diphysa* (Leguminosae). *Acta bot. neerl.* **31,** 135–136.

Holm, L. (1979). Some problems in angiosperm taxonomy in the light of rust data. *Symb. bot. upsal.* **22**(4), 177–181.

Klebahn, H. (1914). Kulturversuche mit Rostpilzen. XV. Bericht (1912 und 1913). *Z. PflKrankh. PflPath. PflSchutz* **24,** 1–32.

Klebahn, H. (1924). Kulturversuche mit Rostpilzen. XVII. Bericht (1916–1924) *Z. PflKrankh. PflPath. PflSchutz* **34,** 289–303.

Leppik, E. E. (1967). Some viewpoints on the phylogeny of rust fungi. VI. Biogenic radiation. *Mycologia* **59,** 568–579.

Mortensen, K. (1986). Biological control of weeds with plant pathogens. *Can. J. Pl. Path.* **8,** 229–231.

Neger, F. W. (1901). Beiträge zur Biologie der Erysipheen. *Flora, Jena* **88,** 333–370.

Neger, F. W. (1902). Beiträge zur Biologie der Erysipheen. II. Mitteilung *Flora, Jena* **90,** 221–272.

Oehrens, E. B. and Gonzalez, S. M. (1974). Introduccion de *Phragmidium violaceum* (Schulz) Winter como factor de control biologico de zarzamora (*Rubus constrictus* Lef. et M. y *R. ulmifolius* Schott). *Agron Sur.* **2,** 30–33.

Oehrens, E. B. and Gonzalez, S. M. (1977). Dispersion, ciclo biologico y danos causados par *Phragmidium violaceum* (Schulz) Winter en zarzamora (*Rubus constrictus* Lef. et M. y *R. ulmifolius* Schott) en la zonas centro-sur y sur de Chile. *Agron Sur.* **5,** 73–85.

Parlevliet, J.E. (1986). Coevolution of host resistance and pathogen virulence: possible implications for taxonomy. In "Coevolution and Systematics" (A. R. Stone and D. L. Hawksworth, eds), pp. 19–34. Clarendon Press, Oxford.

Peterson, R. S. (1973). Studies of *Cronartium* (Uredinales). *Rep. Tottori mycol. Inst.* **10,** 203–223.

Polhill, R. M. and Sousa, M. (1981). Robiniae (Benth.) Hutch 1964. In "Advances in Legume Systematics" (R. M. Polhill and P. H. Raven, eds), pp. 283–288. Royal Botanic Gardens, Kew.

Raven, P. H. and Polhill, R. M. (1981). Biogeography of the Leguminosae. In "Advances in Legume Systematics" (R. M. Polhill and P. H. Raven, eds.), pp. 27–34. Royal Botanic Gardens, Kew.

Savile, D. B. O. (1967). Evolution and relationships of the North American *Pedicularis* rusts and their hosts. *Can. J. Bot.* **45,** 1093–1103.

Savile, D. B. O. (1969). The rusts of *Cheleone* (Scrophulariaceae): a study in coevolution of hosts and parasites. *Nova Hedwigia* **15,** 369–392.

Savile, D. B. O. (1975). Evolution and biogeography of Saxifragaceae with guidance from their rust parasites. *Ann. Mo. bot. Gdn* **62,** 354–361.

Savile, D. B. O. (1979). Fungi as aids in higher plant classification. *Bot. Rev.* **45,** 377–503.

Tranzschel, W. (1939). "Conspectus Uredinalium URSS". Moscow.

Vakili, N. G. and Bromfield, K. R. (1976). *Phakopsora* rust on soybean and other legumes in Puerto Rico. *Pl. Dis. Reptr* **60,** 995–999.

Walker, J. and Stovold, G. E. (1986). *Arkoola nigra* gen. et sp. nov. (Venturiaceae) causing black leaf blight of soybean in Australia. *Trans. Br. mycol. Soc.* **87,** 23–44.

4 Clavicipitaceous Fungal Endophytes of Grasses: Coevolution and the change from parasitism to mutualism

K. CLAY

Department of Biology, Indiana University, Bloomington, USA

Abstract

The mutualistic association between clavicipitaceous fungal endophytes and grasses and its coevolutionary origins are discussed. These endophytic fungi produce an array of alkaloids that are found throughout the aerial plant body and are active against herbivores. Acquired chemical defences against herbivory provide the mechanistic basis for mutualism. Evolution from a localized infection, as seen with *Claviceps*, to a systemic infection, as with *Balansia* and *Epichloë*, results in sexual sterilization of the host plants. Sterile plants, although vegetatively vigorous and resistant to herbivory, can not contribute sexual progeny to future generations. Many host species have been able to overcome this sterility; variation in the degree of endophyte-induced sterility can be seen within several host species and populations. Many fungal endophytes have responded to the restoration of host fertility by losing the ability to reproduce sexually themselves, perfecting a mutualistic association. The loss of sexual reproduction of inhabitants of mutualistic symbioses appears to be common in nature. Chemical defence against herbivory is the basis for mutualism that developed from parasitism by coevolutionary changes in the reproductive systems of host grasses and endophytic fungi.

INTRODUCTION

The relationship between terrestrial plants and fungi is longstanding, perhaps beginning with the initial movement of plants from the sea to land (Pirozynski and Malloch, 1975). Few modern-day vascular plants exist entirely independently of fungi. Many species form mutualistic associations with mycorrhizal fungi from the moment of seed or spore germination; the

COEVOLUTION OF FUNGI
WITH PLANTS AND ANIMALS
ISBN 0-12-557365-0

same plants may be attacked by a variety of parasitic and pathogenic fungi throughout their lives. Coevolutionary changes in the gene pools of host plants and fungi may occur when each species has a major impact on the fitness of the other. Because of the necessity for reciprocal selective pressures, coevolution will be most prevalent between plants and biotrophic fungi that derive their nutrition from living host plants, unlike saprophytic fungi that derive their nutrition from dead plants and plant parts.

The term "coevolution" is used in this contribution in the sense of Janzen (1980), implying an evolutionary change in a trait of individuals in one population in response to a trait of the individuals of a second population, followed by an evolutionary response by the second population to the change in the first population. The range of ecological interactions considered here consequently includes mutualism and parasitism. Mutualism is an interaction between individuals of two species that increases the fitness of both relative to individuals that do not participate in the interaction. In contrast, parasitism is an interaction that enhances the fitness of the organism gaining its nutrition from an individual of the second species, whose fitness is decreased relative to individuals that do not participate in the interaction. The term symbiosis has been used interchangeably with mutualism or to indicate the permanent, physiological integration of individuals of two species without regard to the relative benefit or detriment to each; I follow the usage of de Bary (1879) with no connotation of good or harm. Semantic disagreements over the usage of the terms coevolution, mutualism, and symbiosis are common in the ecological literature (see Lewis, 1985; Pirozynski and Hawksworth, this volume, Chapter 1); it is therefore necessary to define the terms and use them consistently.

Mutualistic interactions between fungi, plants, and animals are widespread. Well-documented examples include mycorrhizas, lichens, fungus-farming ants, and gut microbes in insects (Buchner, 1965; Barrett, 1983; Beattie, 1985; Pirozynski and Hawksworth, this volume, Chapter 1). While fungi have repeatedly formed mutualistic interactions with representatives of the other biological kingdoms, their associations with vascular plants are probably of the greatest economic and ecological significance. The world's terrestrial ecosystems would be very different in the absence of mycorrhizas and lichens. Recently, mutualism between clavicipitaceous fungal endophytes and grasses has been recognized (Siegel *et al.*, 1985; Clay, 1986a). The ecological and agricultural significance of grasses has been found to be due, in part, to this widespread symbiotic association. This paper reviews briefly the relationship between grasses and their fungal endophytes, outlines the evolutionary and coevolutionary processes that took place during the origin of mutualism from parasitism, and indicates how this particular relationship provides insight into coevolution between fungi and vascular plants.

ORIGIN OF MUTUALISM

Before considering the hypothesis that mutualism between grasses and clavicipitaceous fungi has coevolved from parasitism, I consider first how mutualistic associations arise. Janzen (1985) suggests that all mutualisms can be divided into four general categories: pollination, seed dispersal, digestion (defined broadly, including mycorrhizas), or protection.

Parasitism is thought to be the starting point for many of these mutualisms (Boucher *et al.*, 1982; Thompson, 1982). For example, early insects probably fed on pollen, ovules, and seeds of flowering plants; pollination was inadvertant (Crepet, 1979). The highly coevolved relationship between yuccas and moths in the genus *Tegeticula* illustrates how the plants have become dependent for their pollination on a group of insects whose relatives are fruit borers (Aker and Udovic, 1981). Similarly, seed and fruit dispersal by animals probably originated from seed predators that stored their harvest in caches (Janzen, 1971). Considering mycorrhizas, Luttrell (1974) points out that the fungi are found in orders generally characterized by saprotrophy. This is not to say that the mycorrhizal mutualism arose directly from saprotrophism, without passing first through a stage of root parasitism (Harley and Smith, 1983; Barrett, 1983). Pirozynski (1976) and Pirozynski and Malloch (1975) suggested that the very earliest vascular plants existed in association with mycorrhizal fungi; perhaps biotrophy arose from saprotrophy in the sea. The evolutionary origin of fungi and their original role as saprophytes or biotrophs remains to be clearly established (Luttrell, 1974; Barrett, 1983).

The origin of mutualism from a prior parasitic association may differ between symbiotic and non-symbiotic associations. In particular, the association between autotrophic plants and heterotrophic fungal symbionts entails a cost to the plant of providing nutrition for the fungus whether the relationship is mutualistic or parasitic. In non-symbiotic associations there is no necessary cost to the host. One hypothesis for the evolution of mutualism from parasitism is the amelioration model, which states that the detrimental effects of the parasite on its host lessen over evolutionary time until the host benefits from being infected (Boucher *et al.*, 1982). This hypothesis is related to the reduction of parasite virulence; the myxoma virus of rabbits in Australia is a classic example of a virulent parasite becoming avirulent (Fenner and Ratcliffe, 1965). However, amelioration of detrimental effects on host plants is not a sufficient explanation for mutualistic symbioses with heterotrophic fungi (see Figure 1A). Reduced virulence of the parasite over evolutionary time from T_1 to T_3 reduces the differences in fitness of the populations of infected (w_i) and uninfected (w_{ni}) hosts, resulting in benign parasitism. However, the transition to T_4, where the fitness of hosts is greater than non-hosts (mutualism) cannot occur because of the fixed cost of supporting a heterotro-

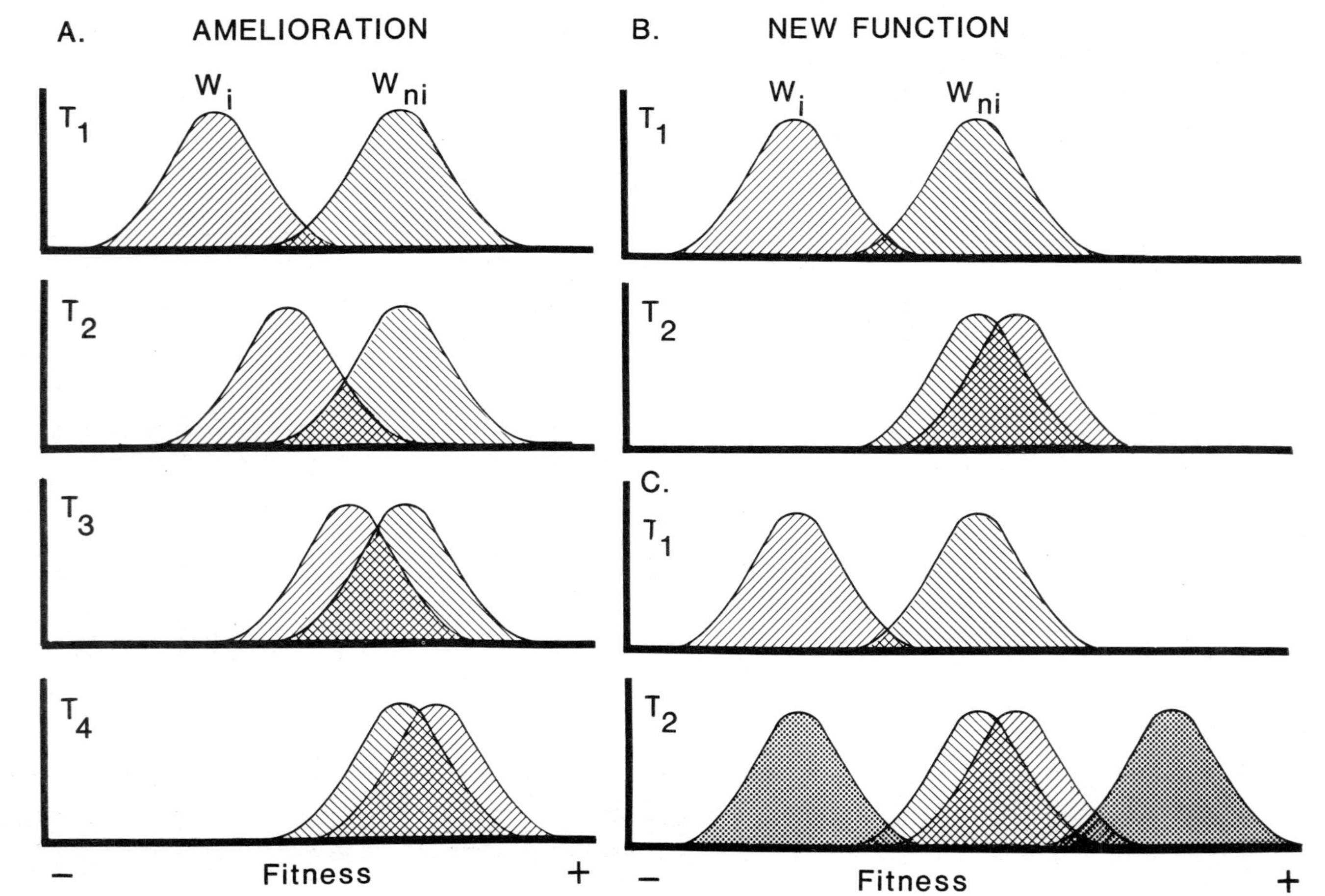
A.
AMELIORATION
W_i
W_{ni}
T_1
T_2
T_3
T_4
−
Fitness
+
B.
NEW FUNCTION
W_i
W_{ni}
T_1
T_2
C.
T_1
T_2
−
Fitness
+

phic symbiont. The evolution of symbiotic mutualism from parasitism must involve some new function of the parasite that results in benefits to the host distinct from the costs of supporting the parasite. This new function may arise from mutation on the part of either partner or from changing environmental conditions, resulting in the sudden appearance of mutualism (Figure 1B). The benefit of the new function, in terms of host fitness, must be greater than the reduction of fitness due to the energetic cost of supporting the fungus (Figure 1C). For example, a relatively avirulent fungus induces resistance of its host to a newly introduced virulent fungus that kills all plants not infected by the first fungus. In most cases of plant–fungal interactions, the various costs and benefits are manifold and their balance is likely to be dynamic; the sum will vary with changing environmental conditions and the genetic identities of the interacting partners. Thus, mutualism and parasitism are not distinct entities; rather they are points along a symbiotic continuum of interacting species.

The persistence and spread of mutualism following its initial appearance depends, in part, on the unit of interaction. If the mutualism is a function of the specific genetic identities of individuals of both species, a situation in which a single plant genotype is interacting with many fungus genotypes (and vice versa) will obscure the expression of mutualism. The majority of plant–fungus genotype combinations will be detrimental to the host. Therefore, the probability that a mutualistic interaction will persist and spread is increased by fidelity of plant and fungus genotype. Fidelity might be difficult to maintain between generations if mutualistic genotype combinations are broken by genetic recombination and dispersal of reproductive propagules. Maternally inherited, seed-borne fungi can maintain compatible genotype combinations through generations and propagate themselves in the gene pools of both species (Clay, 1986a). Such a close genetic association will also tend to favour

---

Figure 1. Two models for the origin of mutualism. **A**, Amelioration Model: Over evolutionary time (T_1 through T_4) the distribution of fitness of hosts (w_i) changes from being much less than that of non-hosts (w_{ni}) at T_1 until it closely approaches that of non-hosts at T_3, resulting in benign parasitism, and finally exceeds the fitness of non-hosts at T_4, resulting in mutualism. **B**, New Function Model: the distribution of fitness of hosts (w_i) changes rapidly, or instantaneously, from less than that of non-hosts (w_{ni}) at T_1 to greater than non-hosts at T_2, resulting in the sudden appearance of mutualism. **C**, is as **B** but the fitness distribution of hosts (w_i) at T_2 is broken down into two components: the cost of supporting the parasite (far left distribution, same as in T_1) and the fitness benefit accruing from the new function (far right distribution). The combined fitness distribution of hosts including the cost and benefit (third distribution from left) exceeds the fitness distribution of non-hosts (second distribution from left).

coevolutionary change; genetic changes in individuals of one species will primarily affect single genotypes of the interacting species.

The potential role of coevolution in mutualistic associations depends on the impact the relationship has on the relative fitnesses of the two partners. A large variance in host fitness due to the presence or absence of a particular fungal partner will result in strong selective pressures favouring the hosts with highest fitness. However, coevolution is a reciprocal process; a large variance in fitness of the fungi also must result from variation in their hosts. For this reason, coevolution will be favoured when there is a high degree of specificity between the interacting partners. For example, some yuccas are pollinated only by yucca moths and the yucca moths feed only on one particular yucca species (Aker and Udovic, 1981). In the case of mycorrhizas, the fungal species may infect the roots of many plant species and, conversely, the plant species may be infected by many fungal species (Barrett, 1983). A genetic change in one fungal species may have little selective impact on a host that is simultaneously infected by many other fungal species. The fitness impact of symbiotic associations is often asymmetric on the two partners. Biotrophic fungi have obligate associations with their partners because of their nutritional requirements; a large proportion of the variance in fitness will necessarily result from genetic differences among hosts for species with restricted host ranges. Coevolution will be most prevalent when host populations are infected by one predominant fungus that has a large influence on plant fitness.

The relationship between several clavicipitaceous fungal endophytes and their host grasses is mutualistic. The family Clavicipitaceae largely comprises parasites of grasses, insects, or other fungi (Jones and Clay, 1987). The fungi are characterized by biotrophy, so it would appear unlikely that the mutualism arose from any other type of association than parasitism. Evidence suggests that the origin of systemic endophytes from fungi with a localized infection, such as *Claviceps*, was associated with herbivore resistance resulting from the production of ergot alkaloids by the fungi present in host tissues. Mutualism originated from parasitism through a series of coevolutionary changes in the reproductive systems of the hosts and fungi, culminating in perfectly fertile hosts infected by asexual, seed-borne, endophytes. Moreover, the association appears to be in a state of transition, with different species combinations found at different positions along the continuum between mutualism and parasitism. This presents an opportunity to examine coevolutionary changes occurring in both host plant and fungus as they move along this continuum. Unlike other associations, in which coevolution is inferred based on historical and comparative evidence, the diversity of endophyte and grass species present now and the variation among populations within species combinations provides a basis for empirical and theoretical studies on the coevolution of plant–fungal interactions. Because many endophytes have only recently been recognized and their significance has not generally been appreciated, the biology of these fungi and their relations with their host grasses is reviewed.

TAXONOMY AND HOST RELATIONS OF CLAVICIPITACEOUS FUNGAL ENDOPHYTES

The endophytic fungi of interest here belong to the tribe Balansiae of the Clavicipitaceae (Ascomycotina) (Diehl, 1950; Latch *et al.*, 1984). The Balansiae includes the genera *Atkinsonella, Balansia, Balansiopsis, Epichloë*, and *Myriogenospora* (Diehl, 1950; Luttrell and Bacon, 1977). The term endophyte is not strictly accurate for a few species known to occur as epiphytes on meristems, young leaves, and inflorescences of their hosts (Luttrell and Bacon, 1977; Leuchtmann and Clay, 1988). However, the term endophyte is used throughout to emphasize the dominant growth habit and phylogenetic relationship of Balansiae species. *Balansia* is the largest genus with about 20 species; the other genera are monospecific or have only a few species. Genera are differentiated primarily by the types of conidia and conidia production (Diehl, 1950). The genera in the tribe Balansiae differ from *Claviceps* in the fruiting bodies (*Claviceps* sclerotia consist entirely of fungal tissue, while in Balansiae genera the fruiting bodies, termed hypothalli or pseudosclerotia by Diehl (1950), consist of both plant and fungal tissues) and the nature of the association between plant and fungus. *Claviceps* is an ovarian parasite of grasses; in contrast, the Balansiae are systemic and perennial in their hosts (Diehl, 1950). While the taxonomy of the genera and species in the Balansiae merits further research, as a tribe the group is clearly distinct from other members of the Clavicipitaceae (Rykard *et al.*, 1984).

Many grasses are asymptomatically infected with endophytes, referred to the anamorph genus *Acremonium* (sect. *Albo-lanosa*), on the basis of their growth characteristics and conidial production in pure culture (Morgan-Jones and Gams, 1982; Latch *et al.*, 1984; White, 1987). These fungi have not been observed to produce the teleomorph, although in other respects they are closely related to or identical to *Epichloë* anamorphs. A number of species have already been described, while other isolates have never yet sporulated in pure culture (Latch *et al.*, 1984; White and Cole, 1985, 1986). From most reported host species, however, the endophytes have never even been cultured. They have, in common with some Balansiae species, growth within leaf sheaths, pith, ovules, and seeds. Elongate, sparsely branched, intercellular hyphae run parallel with the long axis of the plant; no haustoria are produced (Neill, 1941; Siegel *et al.*, 1985). In these respects the endophytes bear little resemblance to other common *Acremonium* species. On the contrary, they clearly have a common phylogenetic origin with *Epichloë* and other Balansiae.

The clavicipitaceous endophytes primarily infect grasses, but two species infect only sedges (Clay, 1986a). In the Balansiae there is specialization within subfamilies of grasses. *Atkinsonella* and *Epichloë* infect C_3, or cool season grasses, while *Balansia*, *Balansiopsis*, and *Myriogenospora* infect C_4, or warm season, grasses (Clay, 1986a). The *Acremonium*-type endophytes primarily infect C_3 grasses, as expected of *Epichloë* anamorphs. The Balansiae fungi produce fruiting bodies that bear both conidia and ascospores on their hosts. Concomitantly, infected plants are sterilized by hormonal inhibition of flowering or mechanical abortion of developing inflorescences (Diehl, 1950; Porter *et al.*, 1985; Clay, 1986a). In contrast, the *Acremonium* endophytes do not fruit on their hosts and the hosts are perfectly fertile. They are transmitted by vegetative growth of the hyphae into ovules and seeds (Clay, 1986a). These coordinated changes in fungus and host reproductive structures represent coevolutionary changes related to the change to mutualism from parasitism.

THE BASIS FOR MUTUALISM—DEFENCE AGAINST HERBIVORY

The Balansiae and their anamorphs are predisposed towards forming mutualistic associations with grasses by virtue of their ability to produce physiologically active alkaloids. The Balansiae and *Claviceps* are largely restricted to grasses that are remarkably free of defensive compounds (Coughenour, 1985). A similar relationship with another plant family (e.g., Leguminosae) would be unlikely because the plants are capable of producing an array of defensive compounds themselves (Janzen, 1971). Anti-herbivore chemistry represents a vacant "niche" in grasses that appears to be filled by symbiotic fungi.

Like *Claviceps*, the endophytes produce a variety of ergot alkaloids in culture and in the host plants (Porter *et al.*, 1979; Bacon *et al.*, 1986; Lyons *et al.*, 1986). All ergot alkaloids have in common the ergoline nucleus, they differ in the structure of attached groups (Figure 2; Cordell, 1981). Three general types of ergot alkaloids are known: clavine alkaloids, lysergic acid alkaloids, and ergopeptide alkaloids. The clavine alkaloids have a $—CH_3$ or $—CH_2OH$ at the C-8 position; lysergic acid alkaloids have a carboxyl or carboxyamide group at the same position; and the ergopeptide alkaloids are amides of lysergic acid with di- or tripeptide attachments (Figure 2; Waller and Dermer, 1981). A single fungal strain can produce several ergot alkaloids and different

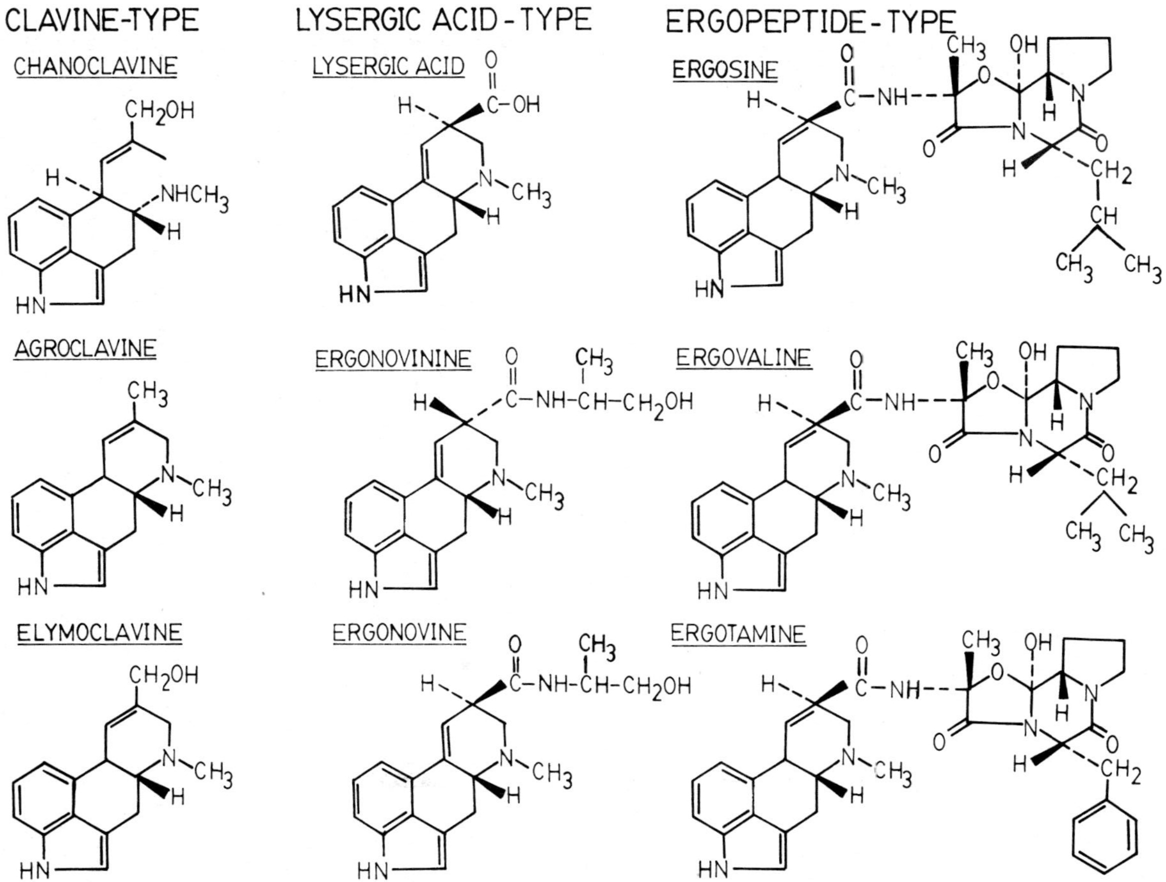

CLAVINE-TYPE

CHANOCLAVINE

AGROCLAVINE

ELYMOCLAVINE

LYSERGIC ACID-TYPE

LYSERGIC ACID

ERGONOVININE

ERGONOVINE

ERGOPEPTIDE-TYPE

ERGOSINE

ERGOVALINE

ERGOTAMINE

Figure 2. Examples of the three major types of ergot alkaloids produced by clavicipitaceous fungal endophytes. All alkaloids illustrated are known to be produced by members of the Balansiae with the possible exception of lysergic acid.

strains produce different arrays of alkaloids (Bacon *et al.*, 1981). Alkaloid production appears to vary within and among strains owing to both environmental and genetic factors. Alkaloids are present in the plant tissues of infected plants but not in uninfected plants (Bush *et al.*, 1982; Hardy *et al.*, 1986; Lyons *et al.*, 1986). The *Acremonium* endophytes also produce several other types of alkaloids in addition to ergot alkaloids. *Acremonium coenophialum* in tall fescue grass (*Festuca arundinacea*) produces pyrrolizidine alkaloids (Jones *et al.*, 1983); lolitrem and peramine alkaloids are produced by *A. lolii* in perennial ryegrass (*Lolium perenne*) (Gallagher *et al.*, 1981; Rowan *et al.*, 1986).

In *Claviceps* the ergot alkaloids are concentrated in the sclerotium (Groger, 1972). Ergot poisonings result from accidental ingestion of sclerotia in grain or forage (Mantle, 1969). In contrast, the alkaloids are systemic in endophyte-infected grasses. Although Hardy *et al.* (1986) and Lyons *et al.* (1986) have shown that alkaloid content varies among plant parts and at different times of the year, they are always present. Therefore, the entire plant and not one particular portion is potentially toxic. The growth habit of the fungi within the plant ensures the widespread distribution of alkaloids in plant tissues. A grass infested with ergot could be less attractive to grazing mammals but an herbivorous insect feeds at a smaller scale and could easily avoid sclerotia. A systemically-infected grass precludes selective feeding of insects on particular plant parts. *Acremonium* endophytes are transmitted through the seed; pyrrolizidine alkaloids have been reported from seeds of endophyte-infected tall fescue (Bush *et al.*, 1982). This raises the possibility that endophyte-infected seeds are less attractive to seed predators.

The closest analogy to the acquired chemical defences found in endophyte-infected grasses are insects that feed on plants and sequester distasteful chemicals, rendering them unpalatable to predators (Gilbert, 1983). It is interesting to note that aposematic butterflies serve as the centrepiece for a dynamic coevolutionary process leading to the formation of mimicry complexes (Gilbert, 1983). The mimics are avoided by predators who associate them with authentic distasteful insects. The common graminoid life-form could even represent mimicry of endophyte-infected grasses.

The relationship between the Balansiae and *Claviceps*, and the long history of ergot poisonings of domestic mammals and man (Mantle, 1969; Groger, 1972), suggests that endophytic fungi could also be poisonous to animals (Diehl, 1950; Cunningham, 1958; Groger, 1972). Early experiments were inconclusive, but recent research has demonstrated that grasses infected by Balansiae and *Acremonium* endophytes are often toxic to mammals (Bacon *et al.*, 1977; Funk *et al.*, 1983; Clay *et al.*, 1985a). Several grasses that have long been known to be toxic (e.g., sleepy grass), have now been shown to be infected by endophytic *Acremonium*-like fungi (Hance, 1876; Bailey, 1903;

McClennan, 1920; White, 1987). *Acremonium coenophialum* and *A. lolii*, infecting tall fescue and perennial ryegrass, respectively, have been linked to large-scale poisonings of cattle and sheep (Bacon *et al.*, 1977; Fletcher and Harvey, 1981; Hoveland *et al.*, 1983). Symptoms include tremors, staggers, increased body temperature, reduced weight gain and milk production, spontaneous abortion, and gangrene of the extremities (Siegel *et al.*, 1987). Afflicted animals recover if quickly moved on to pastures of endophyte-free grasses. Grasses infected by Balansiae endophytes have also been reported to be toxic to domestic mammals (Nobindro, 1934; Bacon *et al.*, 1986). Species of *Balansia* and *Myriogenospora* infecting a variety of grasses give rise to ergot poisoning-like symptoms in cattle, horses, and sheep. It is likely that wild mammalian herbivores also are affected by endophyte-infected grasses, but there are no data available.

Insects feeding on grasses are also affected by the presence of clavicipitaceous endophytes (Table 1). New Zealand perennial ryegrass pastures with a high frequency of endophyte-infected plants had reduced feeding damage and oviposition by the Argentine stem weevil (*Listronotus bonariensis*) (Prestidge *et al.*, 1982). Similarly, Funk *et al.* (1983) in the USA found that sod webworms (*Crambus* spp.) did less feeding damage and laid fewer eggs in stands of infected plants compared with uninfected plants. Subsequently, other insects have been found to discriminate between endophyte-infected and uninfected ryegrass. Tall fescue, the dominant pasture grass in the southeastern USA, when infected by *A. coenophialum* is more resistant to several pests (Johnson *et al.*, 1985; Latch *et al.*, 1985b). We have studied the lepidopteran fall armyworm (*Spodoptera frugiperda*) and several Balansiae-infected grasses and sedges (Clay *et al.*, 1985a,b; Table 1). Host grasses of *Atkinsonella*, *Balansia*, and *Myriogenospora* spp. result in reduced larval survival, reduced rate of weight gain, and increased length of the larval period in comparison with larvae reared on uninfected plants of the same species (Clay *et al.*, 1985a). Two species of sedges infected by *B. cyperi* also resulted in reduced survival and growth of fall armyworm larvae (Clay *et al.*, 1985b). In another study, *Epichloë typhina*-infected *Dactylis glomerata* resulted in reduced survival and weight gain of cutworm larvae (*Agrotis segetus*) (Schmidt, 1986).

To determine whether the presence of endophytes in grass seeds discouraged predation, seeds of perennial ryegrass and tall fescue infected by *Acremonium lolii* and *A. coenophialum*, respectively, were pulverized and supplied to flour beetles (*Tribolium castaneum*). Beetles in vials with endophyte-infected seed had significantly lower initial survival and population growth rate than those supplied with endophyte-free seed (G. P. Cheplick and K. Clay, unpublished). Indeed, the reduced predation of seeds may be of greater adaptive value than the decreased herbivory of vegetative

Table 1. Examples of endophyte-mediated resistance to insect herbivory. Insect responses are in comparison with uninfected plants

Host plant/endophyte and insect response	Reference
Festuca arundinacea infected by *Acremonium coenophialum*	
Aphids *Rhopalosiphum padi* strongly discriminate against infected plants	Latch *et al.* (1985b)
Fall armyworms *Spodoptera frugiperda* have reduced weight gain on infected plants	Clay *et al.* (1985a)
Lolium perenne infected by *A. lolii*	
Crickets *Acheta domesticus* suffer 100% mortality on infected grass	Ahmad *et al.* (1985)
Bluegrass billbugs *Sphenophorus parvulus* do less damage to infected turf	Ahmad *et al.* (1986)
Argentine stem weevils *Listronotus bonariensis* have reduced feeding and oviposition on infected plants and lower densities in highly infected pastures	Barker *et al.* (1984) Mortimer and di Menna (1983)
Sod webworms *Crambus* spp. do less damage to infected turf	Funk *et al.* (1983)
Fall armyworms have reduced survival and weight gain, and increased developmental time on infected plants	Clay *et al.* (1985a)
Dactylis glomerata infected by *Epichloë typhina*	
Cutworms *Agrotis segetum* have reduced survival and weight gain on infected plants	Schmidt (1986)
Paspalum dilatatum infected by *Myriogenospora atramentosa*	
Fall armyworms have reduced survival and weight gain, and increased developmental time on infected plants	Clay *et al.* (1985a)
Cyperus virens infected by *Balansia cyperi*	
Fall armyworms have reduced survival and weight gain, and increased developmental time on infected plants	Clay *et al.* (1985b)

tissues—while a grazed plant quickly grows again, a seed is killed by similar damage.

These data suggest that infected plants are poor food plants compared with uninfected conspecifics. However, insect larvae that take longer to develop ultimately consume more plant tissues, although at greater risk of exposure to predators and pathogens (Scriber and Slansky, 1981). Infected plants, and their fungal symbionts, would be at a greater advantage if insects actively discriminated between infected and uninfected individuals before they fed. The population densities and oviposition frequencies of insects in perennial ryegrass pastures suggests that Argentine stem weevils (*Listronotus bonariensis*) prefer uninfected plants (Prestidge *et al.*, 1982; Mortimer and di Menna, 1983). Laboratory and field experiments confirmed that other insects also differentiated between infected and uninfected perennial ryegrass and tall fescue plants (Barker *et al.*, 1984; Hardy *et al.*, 1985; Johnston *et al.*, 1985; Latch *et al.*, 1985b).

The ability of mammals to detect and avoid endophyte-infected plants has not been as well studied. However, Hance (1876), Shaw (1873) and Bailey (1903) have described how horses and cattle avoided infected species after a single feeding and subsequent poisoning. The behaviour of wild mammalian herbivores awaits further research.

Reduced feeding by herbivorous insects and mammals can result in increased growth and seed production of infected plants, increasing the fitness of both the resident endophyte and host. Pastures of endophyte-infected perennial ryegrass and tall fescue had significantly higher productivity than similar pastures of uninfected plants (Mortimer and di Menna, 1983; Read and Camp, 1986). Others have noted the increased size and vigour of Balansiae-infected plants compared to uninfected individuals in the same population (Diehl, 1950; Bradshaw, 1959; Harberd, 1961; Clay, 1984). We have shown that plants of *Danthonia spicata* infected by *Atkinsonella hypoxylon* were more competitive versus another grass species in a natural grassland community (Kelley and Clay, 1987). Herbivory was not quantified in these studies, so it is not clear whether growth differences resulted from differences in herbivory. Two recent studies suggest, in fact, that increased growth of endophyte-infected plants may be unrelated to herbivory differences. Infected plants of perennial ryegrass and tall fescue grew significantly faster than uninfected plants in controlled environments without herbivory (Latch *et al.*, 1985a; Clay, 1987).

The endophytic fungi are playing a defensive role that benefits both partners. The host plants are subject to reduced herbivory and the fungi enhance their monopoly on their major resource—the living tissues of their host plant. This association may represent a case of diffuse coevolution. The absence of selection for defensive chemistry in grasses is due to the presence of alkaloid-producing fungal endophytes. Similarly, the ubiquitous production of diverse alkaloids by the fungi results from their value in preventing herbivory of their hosts.

COST OF INFECTION—HOST STERILITY

The Balansiae and *Claviceps* are closely related. The greater numbers of *Claviceps* species, their larger host range, and less complex morphologies suggest that the Balansiae were derived from a *Claviceps*-like ancestor. The major evolutionary transition would have been from the localized and transient infection of host plants (as in *Claviceps*) to a systemic and permanent infection (as exemplified by several Balansiae). Walker (1970) has described an unusual systemic *Claviceps* that appears to be intermediate between typical *Claviceps* and the Balansiae. The transition to systemic infection resulted in greater resistance to herbivory, especially by insects, but there was a concomitant cost for the hosts. As indicated by Diehl (1950), most hosts of Balansiae are sterilized by the infection. Inhibition of flowering occurs in some hosts as a result of an alteration of the plant's hormonal balance and (or) the production of phytohormones by the fungi themselves (Porter *et al.*, 1985; Rykard *et al.*, 1985); in other hosts, developing inflorescences are surrounded by fungal tissue, mechanically preventing further development and leading to abortion (Diehl, 1950; Clay, 1986a). *Claviceps* also aborts infected flowers (Bove, 1970; Groger, 1972).

Sterile, infected plants often exhibit increased vegetative growth, perhaps due to herbivore resistance or reallocation of resources. Bradshaw (1959) reported that two species of *Agrostis* infected by *Epichloë typhina* had an increased density of vegetative tillers and spread rhizomatously over wider areas when compared with uninfected plants. Plants of red fescue infected by *E. typhina* were often large compared with uninfected plants (Harberd, 1961), and Diehl (1950) made similar observations on *Cenchrus echinatus* infected by *Balansia obtecta*. Plant size, measured by tiller numbers, was significantly larger for the grass *Danthonia spicata* and the sedge *Cyperus virens* when infected by their respective Balansiae fungi (Clay, 1984, 1986b).

Are vegetatively vigorous, but sterile, hosts advantageous to the fungi? Larger plants typically produce more leaves or more inflorescences than smaller plants. Fungal fruiting bodies (the "pseudosclerotia" or "hypothalli" of Diehl, 1950) are produced on the leaves of hosts inhibited from flowering or on aborted inflorescences. It seems likely that there is a positive correlation between the number of spores produced and host plant size. *Danthonia spicata* and *Cyperus virens* produce aborted inflorescences when infected, and the number of fungal fruiting bodies increases with host plant size (Clay, 1984, 1986b). A second possible advantage of increased vegetative growth of host plants accrues to the fungus: the survival rate of larger plants is greater than that of smaller ones (Clay and Antonovics, 1985). Larger plants are more firmly established and so more buffered from environmental disturbances. The increased host size and survival of infected plants also increases the potential reproductive output and survival of the endophytic fungus.

SEX AND PARASITISM

Levin (1975) suggested that pathogens and parasites provide a selective force favouring the production of genetically variable host offspring. If susceptibility to infection has a genetic basis, as has been demonstrated for many microorganisms, pathogens could readily colonize genetically similar hosts. In contrast, if the offspring were genetically diverse, not all would be equally susceptible to infection. Several authors have postulated related processes by which genetically variable offspring may have higher fitness in the face of pathogen attack (Jaenike, 1978; Hamilton, 1980; Price and Wasser, 1981). Indeed, the role of parasites and pathogens has been a major component in recent experimental and theoretical research on the maintenance of sexual reproduction (Rice, 1983).

If one accepts the logic of these theoretical arguments for the adaptive role of sexual reproduction in host plants and animals faced with parasites and pathogens, the same logic reversed should hold true for the parasites and pathogens themselves. A fungus (or bacterium, etc.) would be selectively favoured that could eliminate sexual reproduction and the production of genetically variable progeny by its host. Offspring that were genetically similar to or identical to the infected parent should be susceptible to the same pathogen. Most Balansiae eliminate or greatly reduce sexual reproduction in their hosts (Diehl, 1950; Clay, 1984, 1986a).

The idea of "parasitic castration" originated in the animal literature (Baudoin, 1975); parasites destroying the reproductive organs of their hosts free energy for increased growth, and castrated hosts exhibit higher survival, increased growth rate, and larger size than comparable unparasitized individuals (Baudoin, 1975). In parallel, the numbers of parasites a host can support, or the size of individual parasites, increases along with the potential numbers of parasite offspring produced by a single host. In this scenario, the advantage to a parasite or pathogen of eliminating the sexuality of its host is ecological rather than genetic. If all individuals in the host population were infected, this strategy would be maladaptive, because eventually the hosts would die and no new members of the host population could be produced. Unlike most animals, many plants have an indeterminate growth form and life span. One individual, while sterile, might grow extensively by vegetative reproduction and live for hundreds or thousands of years (Bradshaw, 1959; Jackson *et al.*, 1985; Wulff, 1985).

Both genetic and ecological arguments have been presented on the possible adaptive value to endophytes of sterilizing their hosts. The genetic argument favouring the elimination of host sexuality by the Balansiae endophytes is less straightforward to document, given the nature of the growth habit in its host. Since the fungi are systemic, growth into reproductive propagules is inherently no different from growth into a new leaf or tiller. If by grafting or somatic mutation there were the physical opportunity for the fungus to grow directly into plant tissues that differed genetically but it could not, this would

Table 2. Distribution of annual, perennial, rhizomatous, and caespitose hosts of *Acremonium*-type and Balansiae endophytes

Host characteristics	Endophyte type		
	Acremonium-like	Balansiae	Total
Perennial, caespitose[a]	31 (82%)	91 (65%)	122
Perennial, rhizomatous	1 (3%)	37 (26%)	38
Annual, caespitose	6 (16%)	12 (9%)	18
Total	38	140	178

[a]G test for null hypothesis that endophyte type is independent of host growth characteristics (perennial hosts only), $G = 12.58$, $P < 0.001$.

symptomless anamorph and as a teleomorph causing partial sterilization of its host. Clay (1986c) also described the symptomless, or latent, infection of *C. rotundus* by *B. cyperi*. These situations are considered probably to represent transitions between Balansiae species causing host sterilization and sterile endophytes producing no symptoms of infection and not impairing host fertility. *Acremonium*-like endophytes probably originated from teleomorphs that once coexisted with latent forms in the same population. This situation is parallel to "secondary" and "primary" species in lichens (Hawksworth, this volume, Chapter 6).

The genetic control of the expression of symptoms of infection on the host and the fruiting of the fungus is not understood. It could be entirely under the control of the fungus, the host, or an interaction of both partners. The variable expression of symptoms of the host and the latency of the fungus suggest that there is underlying genetic control. While experimental cross-innoculations would be the best approach to this question, they are not yet possible with current technical limitations. Another approach is to compare the life-history characteristics of hosts infected by Balansiae causing sterilization with hosts of *Acremonium* endophytes that are completely fertile. Genetic changes in hosts that restore fertility would be strongly selected for in caespitose (clump) grasses but less strongly selected for in rhizomatous or viviparous grasses that have an alternative mode of reproduction. In contrast, genetic changes on the part of the fungus that restored host fertility would be disadvantageous to the fungus by reducing the vegetative vigour of host plants (and the reproductive capacity of the fungus) and allowing the production of genetically variable host progeny that may be unequally susceptible to infection.

In order to determine whether there were such differences in hosts of the *Acremonium* endophytes and of Balansiae, a survey of the literature was conducted (Table 2). White (1987), Funk *et al.* (1983), and Latch *et al.* (1987)

present results of surveys for *Acremonium*-like endophytes in a variety of grasses. Extensive lists of hosts for Balansiae are given in Diehl (1950), Sprague (1950), United States Department of Agriculture (1960), Kohlmeyer and Kohlmeyer (1974), and Rykard *et al.* (1985). Hosts were grouped by subfamily and, following Hitchcock and Chase (1950), classified as being either annual or perennial, and either caespitose or rhizomatous. Thirty-eight species were found to have been recorded as infected by *Acremonium*-like endophytes and classified as to growth habit and lifespan. Over 80% of the hosts are perennial caespitose grasses, whereas only 3% are perennial rhizomatous grasses and 16% are annual grasses. Most host species are from the subfamily Poaidae (syn. Festucoidae), paralleling the distribution of hosts of *E. typhina*. Of 140 hosts recorded for Balansiae, 65% are perennial caespitose grasses, 26% rhizomatous perennials, and 9% annuals (Table 2). Most hosts belong in the subfamily Panicoideae. Thus, a higher percentage of hosts of *Acremonium*-like endophytes are caespitose and annual and a higher percentage of hosts of Balansiae are rhizomatous.

The difference in the distribution of caespitose and rhizomatous perennial hosts of the two types of endophytes is suggestive of different selective pressures resulting from infection. Caespitose hosts infected by Balansiae and rendered sterile have higher survival and growth rates than uninfected conspecifics but have a reduced ability to spread in the population and competitively displace uninfected members. In contrast, rhizomatous or viviparous grasses that are sterilized by infection can still spread vigorously through the population by vegetative means with a lowered probability of genotype extinction (Wulff, 1985). As described by Bradshaw (1959), there are many situations in which grasses grow (heavily grazed sites, marshes, tundra, arctic, etc.) where vegetative reproduction is the predominant mode of reproduction, that is, the cost of sterility is greatly reduced in rhizomatous grasses compared with caespitose grasses. Similarly, the advantage is much greater for caespitose grasses (compared with rhizomatous grasses) when fertility is restored.

A similar argument can be made for the higher frequency of annual grasses as hosts for *Acremonium*-like endophytes. An annual grass sterilized by infection by a fungus of the Balansiae produces neither seed nor vegetative propagules, and the fungus requires high levels of infective spore dispersal in order to persist in a population of annual hosts. One might therefore expect to find a frequent extinction of Balansiae endophytes in annual host populations. In contrast, seed-borne *Acremonium* endophytes entail no reproductive cost for its host and enhance the fungus's probability of finding new hosts. It is perhaps surprising to find any annual hosts for Balansiae.

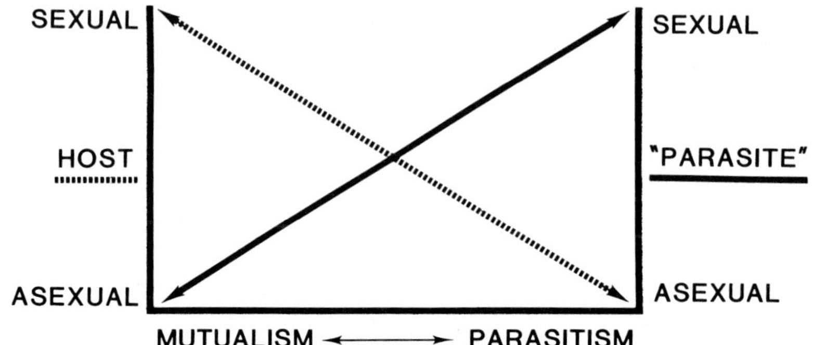

Figure 3. Relationship between reproductive systems of host plant and fungus. In mutualistic associations, the host plant exhibits normal sexual reproduction; the fungus is asexual (*Acremonium*-like). In parasitic associations, the host plant is asexual (sterile) and the fungus is sexual (Balansiae-like). Intermediate situations exist for both plants and fungi (see text).

CHANGES IN ENDOPHYTE REPRODUCTIVE SYSTEMS

There is a continuum of reproductive effects of fungal endophytes on their hosts ranging from complete sterility, as found in most Balansiae hosts, to complete fertility, as seen in tall fescue, ryegrass, and other hosts for *Acremonium* endophytes, with many intermediate conditions as described above. The reproductive systems of the fungi vary in conjunction with the host's reproductive system. The Balansiae causing sterilization of their hosts bear fruiting bodies that produce both ascospores and conidia (Diehl, 1950). In contrast, the *Acremonium* endophytes, which do not cause sterilization of their hosts, do not produce spores on their hosts. Some sporulate in culture, although many do not produce conidia under any circumstances (White and Cole, 1985, 1986) but resemble other *Acremonium* endophytes in other respects. Intermediate expression of sexual reproduction occurs in the fungi as well; *Balansia cyperi* infecting *C. virens* rarely produces ascospores, and on purple nutsedge it has never been observed to produce them (Clay, 1986b,c).

There is an inverse correlation of host and fungus sexuality related to the ecological status of the association: when the host is sexually sterilized, the fungus produces both ascospores and conidia (Figure 3), whereas the fungus is sterile when the host is fertile. Intermediate conditions for both host and fungus have been described above. The relationship between fungus and sterile host must be considered parasitic because, unlike uninfected plants, the host produces no sexual offspring. In contrast, the evidence indicates that the relationship between the *Acremonium*-like endophytes and their hosts, which are fertile, is mutualistic. The hosts are vigorous, resistant to herbivory, and capable of abundant seed production. Thus, in parasitic associations the fungus is sexual and the host is asexual; in mutualistic associations the host is sexual and the fungus is asexual.

THE COEVOLUTIONARY HYPOTHESIS

The ancestral forms of the Balansiae are considered to have resembled extant *Claviceps* in being localized ovarian parasites capable of producing ergot alkaloids. At some point, the fungi became systemic; Walker (1970) has described a possible intermediate state. Host plants were then sterilized by infection, although they were vegetatively vigorous and resistant to herbivory; this is the most common state of Balansiae-infected grasses today. Mutations in host populations that restore fertility would then be favoured (especially in caespitose grasses) by increasing the potential reproductive capacity of plants, in addition to their increased vigour and resistance to herbivores; several examples of Balansiae-infected host grasses capable of producing seed are known (Sampson, 1933; Clay, 1984). The restoration of host fertility would then be followed by the loss of sexual reproduction on the part of the fungi, to become completely internal, seed-borne endophytes as seen in *Acremonium*-like endophytes. Empirical evidence indicates that the relationship between the *Acremonium* endophytes and their hosts is mutualistic; indeed, grasses are known that do not exist in nature free of endophytic fungi (Freeman, 1904; White, 1987; Latch *et al.*, 1987).

Why then should selection favour the abandonment of sexual reproduction by the fungi? Law and Lewis (1983) may provide the answer; they compared a large number of free-living microorganisms that have close relatives living in symbiotic mutualistic associations with plants and animals and found, with few exceptions, that the mutualists exhibited a loss of sexuality compared with their free-living relatives. "The interior of another organism as experienced by a mutualistic inhabitant provides a remarkably constant environment for long periods of time. Once colonized, it generates selective pressures against evolutionary change so that activities such as sex which generate variation are selected against. Such an environment stands in stark contrast with antagonistic ones widespread elsewhere in nature, in which pressures for genetic change are maintained for as long as the organisms continue to exist" (Law and Lewis, 1983, p. 268).

CONCLUSIONS

The association between grasses and clavicipitaceous fungal endophytes provides an example of both coevolution and the change from parasitism to mutualism. The chemical defence of host plants against herbivory by resident fungi is the basis of the mutualism that has generated selective pressures on the reproductive systems of both partners. The idea that grasses benefit from grazing is less tenable when the advantages of being resistant to herbivory are seen in endophyte-infected grasses. Latch *et al.* (1987) suggested that many major pasture grasses (e.g., *Festuca*, *Lolium*, *Poa*) have become endophyte-free only following the development of agricultural techniques of seed harvest and storage. The viability of endophytes in the seeds declines more rapidly than the viability of the seed itself (Neill, 1941; Siegel *et al.*, 1985); extended storage can therefore result in endophyte-free seed (Siegel *et al.*, 1985). Latch *et al.* (1987) found a very high percentage of infected plants in native populations of these grasses in central Europe, but a much lower percentage of infected plants in young populations and recent pastures. If the selective pressures favouring resistance to herbivory were minimal in infected grasses (the fungus providing the defence), they might be expected to be especially palatable and nutritious following the loss of the endophyte, leading to their widespread use in agricultural situations.

The mutualistic symbiosis appears to be quite successful; many grass species are infected, some apparently ubiquitously, and they often dominate grassland communities. The high frequency of infection in certain grass genera (i.e., *Festuca* and *Lolium*) and subfamilies (White and Cole, 1985; White, 1987; Latch *et al.*, 1987) suggests that the relationship is evolutionarily ancient, perhaps originating before the speciation and diversification of ancestral hosts. Indeed, speciation in these groups could have been enhanced by the presence of a mutualistic endophyte. It is somewhat paradoxical that the mutualism, once perfected, can spread only slowly if at all. The *Acremonium* endophytes produce no freely dispersing spores and cannot spread horizontally. Because they are transmitted by the vegetative growth of hyphae into seeds of its host (Clay, 1986a), their increase over time depends entirely on the success of the lineage they infect. Thus, the coevolutionary changes leading to mutualism reduce the potential for further coevolution because of the loss of sexual recombination in the endophytes.

The long-term evolutionary trends in the Gramineae may be intertwined with the evolution of endophytic Clavicipitaceae. Perhaps grasses will eventually all become similar to lichens, obligately infected and dependent on their symbiotic fungi. Certainly, at some point, the Leguminosae existed in an intermediate state with their *Rhizobium* symbionts (Law, 1985). The concept of grasses in the future becoming obligately associated with clavicipitaceous endophytes is not completely improbable, given the advantages that infection by endophytes provides to host plants. If this is true, the current relationship, which is facultative and polymorphic within and among grass species, provides a model in which to observe the evolutionary development of symbiosis between a major group of plants and fungi.

ACKNOWLEDGEMENTS

I thank Drs G. P. Cheplick and A. Leuchtmann, and the editors of this volume for helpful comments on an earlier version of this paper. My research described here has been supported by NSF grants BSR-8400163 and BSR-8614972 and BRSG grant PHS S07 RR 7031J.

REFERENCES

Ahmad, S., Govindarajan, S., Funk, C. R., and Johnson-Cicalese, J. M. (1985). Fatality of house crickets on perennial ryegrasses infected with a fungal endophyte. *Entomologia exp. appl.* **39,** 183–190.

Ahmad, S., Johnson-Cicalese, J. M., Dickson, W. K., and Funk, C. R. (1986). Endophyte-enhanced resistance in perennial ryegrass to the bluegrass billbug, *Sphenophorus parvulus. Entomologia exp. appl.* **41,** 3–10.

Aker, C. L. and Udovic, D. (1981). Oviposition and pollination behavior of the yucca moth, *Tegeticula maculata* (Lepidoptera: Prodoxidae) and its relation to reproductive biology of *Yucca whipplei* (Agavaceae). *Oecologia* **49,** 96–101.

Bacon, C. W., Porter, J. K., Robbins, J. D., and Luttrell, E. S. (1977). *Epichloë typhina* from toxic tall fescue grasses. *Appl. env. Microbiol.* **34,** 576–581.

Bacon, C. W., Porter, J. K., and Robbins, J. D. (1981). Ergot alkaloid biosynthesis by isolates of *Balansia epichloë* and *B. henningsianna. Can. J. Bot.* **59,** 2534–2538.

Bacon, C. W., Lyons, P. C., Porter, J. K., and Robbins, J. D. (1986). Ergot toxicity from endophyte-infected grasses: a review. *Agron. J.* **78,** 106–116.

Bailey, V. (1903). Sleepy grass and its effect on horses. *Science, N.Y.* **17,** 392–393.

Barker, G. M., Pottinger, R. P., Addison, P. J., and Prestidge, R. A. (1984). Effect of *Lolium* endophyte fungus infections on behaviour of adult Argentine stem weevil. *N.Z. Jl agric. Res.* **27,** 271–277.

Barrett, J. A. (1983). Plant-fungus symbioses. In "Coevolution" (D. J. Futuyma and M. Slatkin eds), pp. 137–160. Sinauer Associates, Sunderland, Mass.

de Bary, A. (1879). "Die Erscheinung der Symbiose". Trubner, Strasbourg.

Baudoin, M. (1975). Host castration as a parasitic strategy. *Evolution* **29,** 335–352.

Beattie, A. J. (1985). "The Evolutionary Ecology of Ant-Plant Mutualisms". Cambridge University Press, London.

Boucher, D. H., James, S., and Keeler, K. H. (1982). The ecology of mutualism. *A. Rev. ecol. Syst.* **13,** 315–347.

Bove, F. J. (1970). "The Story of Ergot". Karger, Berlin.

Bradshaw, A. D. (1959). Population differentiation in *Agrostis tenuis* Sibth. II. The incidence and significance of infection by *Epichloë typhina. New Phytol.* **58,** 310–315.

Buchner, P. (1965). "Endosymbiosis of Animals with Plant Microorganisms". Wiley-Interscience, New York.

Bush, L. P., Cornelius, P. L., Buckner, R. C., Varney, D. R., Chapman, R. A., Burrus II, P. B.,

Kennedy, C. W., Jones, T. A., and Saunders, M. J. (1982). Association of *N*-acetyl loline and *N*-formyl loline with *Epichloë typhina* in tall fescue. *Crop Sci.* **22,** 941–943.

Clay, K. (1982). Environmental and genetic determinants of cleistogamy in a natural population of the grass *Danthonia spicata. Evolution* **36,** 734–741.

Clay, K. (1984). The effect of the fungus *Atkinsonella hypoxylon* (Clavicipitaceae) on the reproductive system and demography of the grass *Danthonia spicata. New Phytol.* **98,** 165–175.

Clay, K. (1986a). Grass endophytes. In "Microbiology of the Phyllosphere" (N. J. Fokkema and J. van den Heuvel, eds.), pp. 188–204. Cambridge University Press, London.

Clay, K. (1986b). Induced vivipary in *Cyperus virens*: Transmission of the fungal endophyte *Balansia cyperi. Can. J. Bot.* **64,** 2984–2988.

Clay, K. (1986c). A new disease (*Balansia cyperi*) of purple nutsedge (*Cyperus rotundus*). *Pl. Dis.* **70,** 597–599.

Clay, K. (1987). Effects of fungal endophytes on the seed and seedling biology of *Lolium perenne* and *Festuca arundinacea. Oecologia* **73,** 358–362.

Clay, K. and Antonovics, J. (1985). Demographic genetics of the grass *Danthonia spicata*: Success of progeny from chasmogamous and cleistogamous flowers. *Evolution* **39,** 205–209.

Clay, K. and Jones, J. P. (1984). Transmission of *Atkinsonella hypoxylon* (Clavicipitaceae) by cleistogamous seed of *Danthonia spicata* (Gramineae). *Can. J. Bot.* **62,** 2893–2895.

Clay, K., Hardy, T. N., and Hammond, A. M. Jr. (1985a). Fungal endophytes of grasses and their effects on an insect herbivore. *Oecologia* **66,** 1–6.

Clay, K., Hardy, T. N., and Hammond, A. M. Jr. (1985b). Fungal endophytes of *Cyperus* and their effect on an insect herbivore. *Am. J. Bot.* **72,** 1284–1289.

Cordell, G. A. (1981). "Introduction to Alkaloids: A Biogenic Approach". Wiley, New York.

Coughenour, M. B. (1985). Graminoid responses to grazing by large herbivores: adaptations, exaptations, and interacting processes. *Ann. Mo. Bot. Gdn* **72,** 852–863.

Crepet, W. L. (1979). Insect pollination: a paleontological perspective. *Bioscience* **29,** 102–108.

Cunningham, J. J. (1958). Non-toxicity to animals of ryegrass endophyte and other endophytic fungi of New Zealand grasses. *N.Z. Jl agric. Res.* **1,** 489–497.

Diehl, W. W. (1950). "*Balansia* and the Balansiae in America". (Agriculture Monograph no. 4.) United States Department of Agriculture, Washington D.C.

Fenner, F. and Ratcliffe, F. N. (1965). "Myxomatosis". Cambridge University Press, London.

Fletcher, L. R. and Harvey, I. C. (1981). An association of a *Lolium* endophyte with ryegrass staggers. *N.Z. Vet. J.* **29,** 185–186.

Freeman, E. M. (1904). The seed fungus of *Lolium tementulum* L., the darnel. *Phil. Trans. Soc., B* **214,** 1–28.

Funk, C. R., Halisky, P. M., Johnson, M. C., Siegel, M. R., Stewart, A. V., Ahmad, S., Hurley, R. H., and Harvey, I. C. (1983). An endophytic fungus and resistance to sod webworms: association in *Lolium perenne* L. *Bio/Technology* **1,** 189–191.

Gallagher, R. T., White, E. P., and Mortimer, P. H. (1981). Ryegrass staggers: isolation of potent neurotoxins lolitrem A and lolitrem B from staggers producing pastures. *N.Z. Vet. J.* **29,** 189–190.

Gilbert, L. E. (1983). Coevolution and mimicry. In "Coevolution" (D. J. Futuyma and M. Slatkin, eds.), pp. 263–281. Sinauer Associates, Sunderland, Mass.

Groger, D. (1972). Ergot. In "Microbial Toxins" (S. Kadis, A. Ciegler, and S. J. Ajl, eds), vol. 7, pp. 321–373. Academic Press, New York.

Hamilton, W. D. (1980). Sex versus non-sex versus parasite. *Oikos* **35,** 282–290.

Hance, H. F. (1876). On a Mongolian grass producing intoxication in cattle. *J. Bot., Lond.* **14,** 210–212.

Harberd, D. J. (1961). Note on choke disease of *Festuca rubra. Rep. Scott. Pl. Breed. Stn* **1961,** 47–51.

Hardy, T. N., Clay, K., and Hammond, A. M. (1985). Fall armyworm (Lepidoptera: Noctuidae): A laboratory bioassay and larval preference study for the fungal endophyte of perennial ryegrass. *J. econ. Ent.* **78,** 571–575.

Hardy, T. N., Clay, K., and Hammond, A. M. (1986). The effect of leaf age and related factors on endophyte-mediated resistance to fall armyworm (Lepidoptera: Noctuidae) in tall fescue. *J. env. Ent.* **15,** 1083–1089.

Harley, J. H. and Smith, S. E. (1983). "Mycorrhizal Symbiosis". Academic Press, London.

Hitchcock, A. S. and Chase, A. (1950). "Manual of the Grasses of the United States". United States Government Printing Office, Washington D.C.

Hoveland, C. S., Schmidt, S. P., King, C. C., Odum, J. W., Clark, E. M., McGuire, J. A., Smith, L. A., Grimes, H. W., and Holliman, J. L. (1983). Steer performance and association of *Acremonium coenophialum* fungal endophyte on tall fescue pasture. *Agron. J.* **75,** 821–824.

Jackson, J. B. C., Buss, L. W., and Cook, R. E. (1985). "Population Biology and Evolution of Clonal Organisms". Yale University Press, New Haven.

Jaenike, J. (1978). An hypothesis to account for the maintenance of sex within populations. *Evol. Theor.* **3,** 191–194.

Janzen, D. H. (1971). Seed predation by animals. *A. Rev. ecol. Syst.* **2,** 465–492.

Janzen, D. H. (1980). When is it coevolution? *Evolution* **34,** 611–612.

Janzen, D. H. (1985). The natural history of mutualisms. In "The Biology of Mutualism" (D. H. Boucher, ed.), pp. 40–99. Oxford University Press, New York.

Johnson, M. C., Dahlman, D. L., Siegel, M. R., Bush, L. P., Latch, G. C. M., Potter, D. A., and Varney, D. R. (1985). Insect feeding deterrents in endophyte-infected tall fescue. *Appl. env. Microbiol.* **49,** 568–571.

Jones, J. P. and Clay, K. (1987). Ascus and crozier development in the Balansiae. *Can. J. Bot.* **65,** 1027–1030.

Jones, T. A., Buckner, R. C., Burrus, P. B. II and Bush, L. P. (1983). Accumulation of pyrrolizidine alkaloids in benomyl-treated tall fescue parents and their untreated progenies. *Crop Sci.* **23,** 1135–1140.

Kelley, S. E. and Clay, K. (1987). Interspecific competitive interactions and the maintenance of genotypic variation within the populations of two perennial grasses. *Evolution* **41,** 92–103.

Kohlmeyer, J. and Kohlmeyer, E. (1974). Distribution of *Epichloë typhina* (Ascomycetes) and its parasitic fly. *Mycologia* **66,** 77–86.

Latch, G. C. M., Potter, L. R. and Tyler, B. F. (1987). Incidence of endophytes in seeds from collections of *Lolium* and *Festuca* species. *Ann. appl. Biol.* **111**, 59–64.

Latch, G. C. M., Christensen, M. J., and Samuels, G. J. (1984). Five endophytes of *Lolium* and *Festuca* in New Zealand. *Mycotaxon* **20,** 535–550.

Latch, G. C. M., Hunt, W. F., and Musgrave, D. R. (1985a). Endophytic fungi affect growth of perennial ryegrass. *N.Z. Jl agric. Res.* **28,** 165–168.

Latch, G. C. M., Christensen, M. J., and Gaynor, D. L. (1985b). Aphid detection of endophyte infection in tall fescue. *N.Z. Jl agric. Res.* **28,** 129–132.

Law, R. (1985). Evolution in a mutualistic environment. In "The Biology of Mutualism" (D. H. Boucher, ed.), pp. 145–170. Oxford University Press, New York.

Law, R. and Lewis, D. H. (1983). Biotic environments and the maintenance of sex—some evidence from mutualistic symbioses. *Biol. J. Linn. Soc.* **20,** 249–276.

Levin, D. A. (1975). Pest pressure and recombination systems in plants. *Am. Nat.* **109,** 437–451.

Lewis, D. H (1985). Symbiosis and mutualism: crisp concepts and soggy semantics. In "The Biology of Mutualism" (D. H. Boucher, ed.), pp. 29–39. Oxford University Press, New York.

Leuchtmann, A. and Clay, K. (1988). *Atkinsonella hypoxylon* and *Balansia cyperi*, epiphytic members of the Balansiae. *Mycologia* **80,** 192–199.

Luttrell, E. S. (1974). Parasitism of fungi on vascular plants. *Mycologia* **66,** 1–15.

Luttrell, E. S. and Bacon, C. W. (1977). Classification of *Myriogenospora* in the Clavicipitaceae. *Can. J. Bot.* **55,** 2090–2097.

Lyons, P. C., Plattner, R. D., and Bacon, C. W. (1986). Occurrence of peptide and clavine ergot alkaloids in tall fescue grass. *Science, N.Y.* **232,** 487–489.

Mantle, P. G. (1969). The role of alkaloids in the poisoning of mammals by sclerotia of *Claviceps* spp. *J. stored Products Res.* **5,** 237–244.

McClennan, E. (1920). The endophytic fungus of *Lolium*, Part I. *Proc. R. Soc. Vict.* **11,** 252–301.

Morgan-Jones, G. and Gams, W. (1982). Notes on hyphomycetes. XLI. An endophyte of *Festuca arundinacea* and the anamorph of *Epichloë typhina*, new taxa in one of the two new sections of *Acremonium. Mycotaxon* **15,** 311–318.

Mortimer, P. H. and di Menna, M. E. (1983). Ryegrass staggers: further substantiation of a *Lolium* endophyte aetiology and the discovery of weevil resistance of ryegrass pastures infected with *Lolium* endophyte. *Proc. N.Z. Grassl. Assoc.* **44,** 240–243.

Neill, J. C. (1941). The endophytes of *Lolium* and *Festuca. N.Z. Jl Sci. Tech.* **23,** 185–193.

Nobindro, U. (1934). Grass poisoning among cattle and goats in Assam. *Indian Vet. J.* **10,** 235–236.

Philipson, M. N. and Christey, M. C. (1985). An epiphytic/endophytic fungal associate of *Danthonia spicata* transmitted through the embryo sac. *Bot. Gaz.* **146,** 70–81.

Pirozynski, K. A. (1976). Fossil fungi. *A. Rev. Phytopath.* **14,** 237–246.

Pirozynski, K. A. and Malloch, D. W. (1975). The origin of land plants: a matter of mycotrophism. *BioSystems* **6,** 153–164.

Porter, J. K., Bacon, C. W., and Robbins, J. D. (1979). Ergosine, ergosinine, and chanoclavine I from *Epichloë typhina. J. agric. Fd Chem.* **27,** 595–598.

Porter, J. K., Bacon, C. W., Cutler, H. G., Arrendale, R. F., and Robbins, J. D. (1985). *In vitro* auxin production by *Balansia epichloë. Phytochemistry* **24,** 1429–1431.

Prestidge, R. A., Pottinger, R. P., and Barker, G. M. (1982). An association of *Lolium* endophyte with ryegrass resistance to Argentine stem weevil. *Proc. N.Z. Weed Pest Control Conf.* **35,** 199–222.

Price, M. V. and Wasser, N. M. (1981). Population structure, frequency dependent selection, and the maintenance of sexual reproduction. *Evolution* **36,** 35–43.

Read, J. C. and Camp, B. J. (1986). The effect of fungal endophyte *Acremonium coenophialum* in tall fescue on animal performance, toxicity, and stand maintenance. *Agron. J.* **78,** 858–860.

Rice, W. R. (1983). Parent-offspring pathogen transmission: a selective agent promoting sexual recombination. *Am. Nat.* **121,** 187–203.

Rowan, D. D., Hunt, M. B., and Gaynor, D. L. (1986). Peramine, a novel insect feeding deterrent from ryegrass infected with the endophyte *Acremonium loliae. J. chem. Soc., Chem. Comm.* **142,** 935–936.

Rykard, D. M., Luttrell, E. S., and Bacon, C. W. (1984). Conidiogenesis and conidiomata in the Clavicipitoideae. *Mycologia* **76,** 1095–1103.

Rykard, D. M., Bacon, C. W., and Luttrell, E. S. (1985). Host relations of *Myriogenospora atramentosa* and *Balansia epichloe* (Clavicipitaceae). *Phytopathology* **75,** 950–956.

Sampson, K. (1933). The systemic infection of grasses by *Epichloë typhina* (Pers.) Tul., *Trans. Br. mycol. Soc.* **18,** 30–47.

Schmidt, D. (1986). La quenouille rend-elle le fourrage toxique? *Revue suisse Agric.* **18,** 329–332.

Scriber, J. M. and Slansky, F. (1981). The nutritional ecology of immature insects. *A. Rev. Entom.* **26,** 183–211.

Shaw, J. (1873). On the changes going on in the vegetation of South Africa. *J. Linn. Soc., Bot.* **14,** 202–208.

Siegel, M. C., Latch, G. C. M., and Johnson, M. C. (1985). *Acremonium* fungal endophytes of tall fescue and perennial ryegrass: significance and control. *Pl. Dis.* **69,** 79–83.

Siegel, M. C., Latch, G. C. M., and Johnson, M. C. (1987). Fungal endophytes of grasses. *A. Rev. Phytopath.* **25,** 293–315.

Sprague, R. (1950). "Diseases of Cereals and Grasses (Fungi, except Smuts and Rusts)". Ronald Press, New York.

Stovall, M. E. and Clay, K. (1988). The effect of the fungus *Balansia cyperi* on growth and reproduction of purple nutsedge, *Cyperus rotundus*. *New Phytol.*, in press.

Thompson, J. N. (1982). "Interaction and Coevolution". Wiley, New York.

United States Department of Agriculture (1960). "Index of Plant Diseases in the United States". [USDA Agricultural Handbook no. 165.] United States Government Printing Office, Washington D.C.

Walker, J. (1970). Systematic fungal parasite of *Phalaris tuberosa* in Australia. *Search* **1,** 81–83.

Waller, G. R. and Dermer, O. C. (1981). Enzymology of alkaloid metabolism in plants and microorganisms. In "The Biochemistry of Plants" (E. E. Conn, ed.), vol. 7, pp. 317–402. Academic Press, New York.

White, J. F. (1987). The widespread distribution of endophytes in the Poaceae. *Pl. Dis.* **71,** 340–342.

White, J. F. and Cole, G. T. (1985). Endophyte-host associations in forage grasses. I. Distribution of fungal endophytes in some species of *Lolium* and *Festuca*. *Mycologia* **77,** 323–327.

White, J. F. and Cole, G. T. (1986). Endophyte-host associations in forage grasses. V. Occurrence of fungal endophytes in certain species of *Bromus* and *Poa*. *Mycologia* **78,** 852–856.

Wulff, J. L. (1985). Clonal organisms and the evolution of mutualism. In "Population Biology and Evolution of Clonal Organisms" (J. B. C. Jackson, L. W. Buss, and R. E. Cook, eds), pp. 437–466. Yale University Press, New Haven, Conn.

5 | Observations on the Coevolution of Fungi with Hepatics

B. BOULLARD

Biologie Végétale, Université de Rouen, France

Abstract

A brief survey of fossil Hepaticae, of the hypothetical ancestors of the group, and of the main lines of the evolution of the liverworts, is followed by consideration of frequent symbiosis of their gametophytes with fungi (here called "mycothalli"). The systematic distribution of mycothallic Hepaticae, their morphology, and data pertaining to the fungi involved, suggest that coevolution with symbiotic fungi has been a significant factor in the evolutionary history of the Hepaticae.

INTRODUCTION

Today, no biologist denies the fact of evolution. Here and there throughout geological time, two groups of living organisms have been able to establish more or less intimate relationships and have so blended together their evolution; we refer to such cases as coevolution. Pteridophytes and their symbiotic fungi offer a clear example (Boullard, 1979), and others are considered elsewhere in this volume. The hepatics are a further instance treated here.

Alongside mosses and hornworts, the liverworts constitute an important group of bryophytes, distinguished by a long-lived haplophasic state (excepting the gametophytes of Sphaerocarpales; Schuster, 1984, p. 1039) as opposed to a brief diploid and epiphytic phase. Haploid organisms have been considered less resistant to environmental change than diploid ones (Schuster, 1983, p. 138), and gametophytes of hepatics appear to be excessively plastic. No group of plants with a similar number of extant species has such diversity and exhibits so many evolutionary peculiarities (Schuster, 1984, p. 892). Consequently, in this group, any phylogenetic approach remains an "attempt", so common are equivocal similarities. Data provided by chemo-

COEVOLUTION OF FUNGI
WITH PLANTS AND ANIMALS
ISBN 0-12-557365-0

taxonomy are not always helpful because "liverworts can exist in different chemical races (chemotypes) or geographical races" (Schuster, 1983, p. 41). Nevertheless, *Monoselenium*, for instance, can be distinguished from the Marchantiaceae on the basis of the flavonoid content (Campbell *et al.*, 1979).

As palaeobotanical data (see below) are too scarce to permit a reliable reconstruction of the phylogeny of the Hepaticae, current generalizations concerning phylogenetic relationships (e.g., Mehra, 1968, 1969; Schuster, 1984) can be expected to require revision in future.

FOSSIL HEPATICAE

Fossil Hepaticae are very scarce prior to the Cenozoic (Fulford, 1965) and contribute little to clarifying the evolution of the group because *Riccia*, *Marchantia* and the Jungermanniales are geologically more ancient. By the end of the Paleozoic, or the start of the Mesozoic, the diversity of present-day phyla had already been reached. The main systematic units (Metzgeriales, Sphaerocarpales, Jungermanniales, Marchantiales) probably appeared as early as the first quarter of the Mesozoic. No fossil Takakiales or Calobryales have been found, but the oldest Metzgeriales existed in the Paleozoic, before the advent of Jungermanniales, and particularly the still more recent Marchantiales.

The oldest bryophytes were undoubtedly "contemporaneous with early vascular plants" (Crandall-Stotler, 1984, p. 1096) that is, with the late Silurian/early Devonian *Cooksonia*, *Horneophyton* and *Rhynia*. In view of the lack of sufficient fossil material, we are obliged "to draw conclusions from ontogeny, comparative morphology, anatomy, and from analogies with other groups of terrestrial plants" (Evans, 1939, p. 59).

The distinctiveness of mosses, liverworts, and hornworts appears to be very ancient, dating from the time of their origin according to Crandall-Stotler, 1984, p. 1108), who considered the morphogenetic designs of these groups. Schuster (1984, p. 913) believed that the "subclasses and orders of hepatics were established very early, surely by Carboniferous times, and possibly by Devonian times". On the basis of unquestionable Paleozoic fossils, we can be confident of the existence of different families of Metzgeriales in that time. Moreover, if *Pallaviciniites devonicus* truly belongs to the Metzgeriales (Schuster, 1984, p. 45), then the Metzgeriales and Jungermanniales had already diverged by the lowermost Upper Devonian. This agrees with the postulated late Silurian/early Devonian origin of the Hepaticae. Their ancestors should be searched for in deposits from these geological periods, even if there is "no general consensus as to what this ancestral type of liverwort was like" (Schuster, 1984, p. 893).

ANCESTRAL HEPATICAE

To find surviving relics of the more ancestral liverworts, the old area of Gondwanaland is most promising in as much as "remnants of this ancient Gondwanalandic flora persisted in New Zealand, New Caledonia . . ." (Schuster, 1983, p. 493). Moreover, to assist in determining the morphology of the common ancestor of the group, it may be necessary to consider "a composite picture": it will be surprisingly similar to *Takakia*, and even to *Haplomitrium* (e.g., *H. intermedium* or *H. ovalifolium*), both genera strongly suggesting "that we do have a logical starting point for some, if not all, hepatic evolution" (Schuster, 1984, p. 46).

Crandall-Stotler (1984, p. 109) considered the progenitors of bryophytes to have had isomorphic alternating life-forms. In support of this opinion the following be noted. (a) Plants with isomorphic life cycles include algae that provide a link with chlorophyllous but still non-vascular plants. (b) Chlorophyllous autotrophic sporophytes are well-suited for independence as gametophytes and exist in various groups of bryophytes (e.g., *Cephalozia, Pellia, Marchantia, Dumortiera, Ricciocarpus*). (c) A number of primitive pteridophytes clearly support the isomorphic (or homologous) theory (e.g., *Rhynia, Psilotum, Tmesipteris, Lycopodium, Ophioglossum* p.p.; Boullard, 1979).

The epiphytic position of the sporophyte, permanently supported by the gametophyte, hinders aquatic fertilization in bryophytes. Prostrate gametophytes (and *a fortiori*, subterranean and mycothallic ones) differentiated very early in pteridophyte evolution (cf. Boullard, 1979). Indeed, it has to be considered whether pteridophytes and bryophytes evolved from a biologically similar mycotrophic ancestor. Bernard (1909, p. 18) had already supposed that ancestral vascular plants could have arisen, through adaptation to mutualistic symbiosis and subsequent vascularization, from some extinct liverwort with a mycothallic and long-lived gametophyte. Magrou (1948) accepted this hypothesis, claiming parallels between *Pellia epiphylla* and *Horneophyton lignieri*; unfortunately we still cannot say whether the tuberous and moniliform base of the latter is gametophytic.

If the ancestor of the Hepaticeae ever possessed a symbiotic gametophyte, bearing an epiphytic sporophyte, a clear relationship could be seen with *Rhynia**, and even *Horneophyton* (despite the uncertainty mentioned above), or *Cooksonia* (of which, unfortunately, the lower part has not yet been discovered). Antithetic alternation with haploid dominance, so typical in bryophytes, would have arisen later.

---

*According to Edwards (1986), *Rhynia major* is not a tracheophyte but a plant related to the Bryophyta. This concept is consistent with the ideas advocated in this contribution concerning the "archetype" of the Hepaticae and the antiquity of their symbiosis with fungi.

GAMETOPHYTIC FUNGAL SYMBIOSIS*

In mycotrophic flowering plants, only the diploid phase is able to support symbionts; we speak of such associations, and of similar associations involving sporophytes of numerous Pteridophytes, as mycorrhizas or mycorrhizomes, according to the nature of the infected organ. On the other hand, symbiotic gametophytes (i.e., thalli) of some Pteridophytes and Hepaticae can be referred to as "mycothalli" (Boullard, 1979, p. 11).

The environmental conditions (e.g., soil type, humus and water contents, light, temperature) markedly affect the intensity of mycotrophy in plants. Consequently the adaptation of Marchantiales to grow in intense light cannot be isolated from their conspicuous mycotrophy, considering the adaptations of mycotrophic plants as a whole (Peyronel, 1939; Boullard, 1960).

Likewise, very wet conditions appear to be unfavourable to fungal symbiosis. Marchantiidae, which arose by the early Mesozoic, had to resist desiccation just as they do today by having xeromorphic thalli, while those of Monocleales or Sphaerocarpales are drought-intolerant. The frequent occurrence of mycothalli in Marchantiaceae was therefore to be expected. In these gametophytes (see below) the mycobiont occupies a specific band-shaped zone, often brownish in colour, as in *Neohodgsonia* and *Preissia* (Figure 1C).

The nearly cosmopolitan geographical distributions of *Reboulia*, *Aneura*, *Metzgeria leptoneura* or *Dumortiera hirsuta* offer interesting possibilities for research into symbiotic behaviour in different locations. *Riccia*, which shows broad ecological diversity, or Jungermanniales, which are strongly affected by environmental stresses, would be prime candidates for the investigation of correlations between symbiotic status and habitat.

As for mutualistic symbiosis with fungi facilitating the survival of plants growing under severe conditions, I agree with Mehra (1968, p. 55) that "there is nothing like a wholesale progression of reduction. Nature knows no such thing, what it knows is the adaptability of the living forms to their environments for survival."

Mycothalli and systematics of Hepaticae

Assuming that the gametophyte always supported the sporophyte, potential mycotrophy in hepatics primarily concerns the former. The habit and habitat of hepatics (where thalli and rhizoids are closely associated with the substratum) makes them particularly vulnerable to infection by fungal mycelia.

*Increasingly, biologists adopt the original de Baryan concept of "symbiosis" which is very broad, encompassing all cases of "living together" of different organisms (see Pirozynski and Hawksworth, Chapter 1). This chapter focuses on mutualistic rather than commensalistic or parasitic relationships, that is, "mycothalli".

However, this intimate contact with soil has not always been realized, owing to the repeated independent evolution of erect or prostrate and flattened habits. It is, therefore, possible that mycotrophy occurred only at intervals in suitable conditions.

Unfortunately, fungal symbionts have not yet been detected in undisputed fossil Hepaticae. From Table 1 it is clear that mycotrophy is not the prerogative of apparently primitive taxa. Thus, in the Lepidoziaceae, the species of *Lepidozia* are considered to be very old, and the species of *Zoopsis* to be undoubtedly advanced; both build mycothalli. And, as Pocock and Duckett (1985b) observed, "Fungi are absent from the most primitive and most advanced families in Jungermanniales (Herbertaceae and Lejeuneaceae respectively)". I also agree with these authors that "fungi tend to be absent from ephemerals, epiphytes, and species growing in very wet habitats"; this last factor has been referred to above.

Morphology of Mycothalli

Gametophytes of the Hepaticae are phenomenally malleable (Schuster, 1984, p. 762). Although it seems that spores can germinate in the absence of the fungi, subsequent infection always results in new morphogenetic features in the host (e.g., branching or swelling of rhizoids, localized or general fleshiness of the thallus). However, apical ramification of rhizoids sometimes appears independently of fungal symbiosis.

Occasionally, as in *Conocephalum conicum* (Bolleter, 1905), infected thalli show simultaneous reduction in length and a tendency to fleshiness. Sometimes tubers are formed. Apart from their xerophytic adaptation (cf. Chalaud, 1932), it would be useful to know whether such tubers also depend upon an efficient mutualistic symbiosis.

The presence and the localization of fungal hyphae is, for various taxa, as typical as it is constant (e.g., in Lophoziaceae). Considering that the main phyla of the Hepaticae were present as far back as the Devonian/Carboniferous, similar biological behaviour of mycothalli must have evolved repeatedly in response to comparable environmental factors. Some aspects of individual host resistance influence the synthesis of mycothalli or the intimacy of integration.

Sometimes infection is limited to the rhizoids, the swollen apices of which become filled with the endophyte (Figure 2G–I). Often, "these branched and swollen rhizoids are associated with leafless and largely achlorophyllous flagellar branches" (Pocock *et al.*, 1984b) that simulate "root systems", as in *Takakia* or *Haplomitrium*. Hepaticae with swollen rhizoidal apices are characteristic of trophically peculiar microhabitats such as strongly oligotrophic

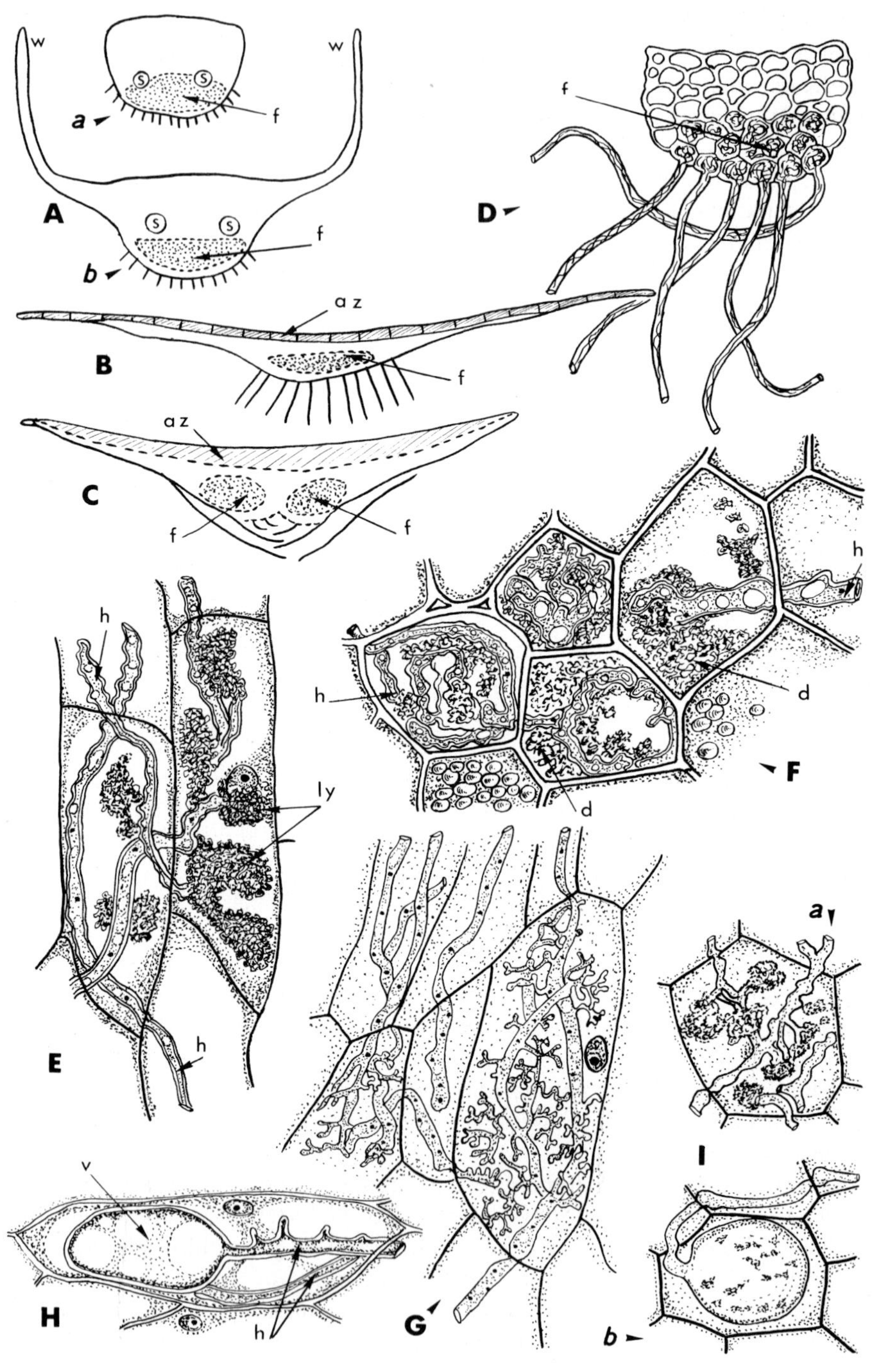

w
w
S
S
f
a
A
S
S
f
b
D
f
a z
B
f
a z
C
f
f
h
h
d
h
d
F
ly
h
E
a
G
I
v
h
H
b

bogs. Pocock *et al.* (1984) suggested that "the fungi have an important absorptive role" (see below).

In thalloid species, especially Marchantiales, the endophytic fungus commonly occupies an important axial zone, or even a large median band extending into the wings of the thallus (Figure 1A–C).

The mutualistic symbiotic associations in Hepaticeae comprise the following morphological types (Burgeff, 1938; Stahl, 1949; Table 2):

(a) vesicular-arbuscular (VA) mycothalli identical with the ubiquitous VA mycorrhizae of the phanerogams ("thamniscophagous" *sensu* Burgeff, 1938); see Figure 1E–I;
(b) vesicular-arbuscular mycothalli, but with a particularly prolific production of vesicles, which disintegrate together with arbuscules ("thamniscophysalidophagous" *sensu* Stahl, 1949); see Figure 2A–B;
(c) mycothalli with intracellular, coiled, generally septate hyphae ("pelotons"), and clumped bodies generated by digestion of coils ("tolypophagous" *sensu* Burgeff, 1938); see Figures 1D, and 2C–D;
(d) mycothalli with rhizoids harbouring fungal hyphae, sometimes grouped into basal pseudoparenchymatous "tissue" and generating haustoria (to be digested later) in adjacent cells ("haplohaustorial" *sensu* Stahl, 1949); see Figure 2E–F;
(e) in addition, Pocock and Duckett (1984a) discovered mycothalli in which infection is almost totally limited to swollen apices of rhizoids, which become filled with hyphae (Figure 2G–I).

The fungi involved

It is clear that little is known about the identity of fungal mutualists of liver-

Figure 1. A, *Möerckia flotowiana*, transverse sections through basal (1a) and medial (1b) parts of the plant; s, conducting strands; f, fungal zones; w, wings of the thallus (after Cavers, 1911). **B**, *Pellia epiphylla*, transverse section of thallus; f, fungal zone; az, air chambers zone. **C**, *Preissia commutata*, transverse section of thallus; f, fungal zones; az, air chambers zone (after Cavers, 1911). **D**, *Lophozia bicrenata*, cross section of a stem showing fungal infection (f) in its lower part and rhizoids. **E**, *Takakia lepidozioides*, longitudinal section of the stem; intracellular hyphae (h) in early stages of lysis (ly). Thamniscophagous mycothallus. **F**, *Pellia epiphylla*; intact (h) and digested (d) hyphae. Thamniscophagous mycothallus. **G**, *Lunularia cruciata*, arbuscules. Thamniscophagous mycothallus (after Auret, 1930). **H**, *Preissia commutata*, intracellular hyphae (h) and a vesicle (v) with oil bodies. Thamniscophagous mycothallus (after Golenkin, 1902). **I**, *Sewardiella tuberifera*: (a) lysis of arbuscule and formation of ptyosomes; (b) intracellular vesicle. Thamniscophagous mycothallus (after Chalaud, 1932).

Table 1. Systematic distribution of mycotrophism in the Hepaticeae (+, mycothalli present; −, mycothalli absent)

Order	Family	Association with fungi				
		Regular or frequent		Occasional		No data
		+	−	+	−	
Calobryales	Takakiaceae	*				
	Haplomitriaceae	*				
Metzgeriales	Fossombroniaceae	*				
	Pelliaceae	*				
	Allisoniaceae					*
	Phyllothalliaceae					*
	Pallaviciniaceae	*			*	
	Sandeothallaceae					*
	Makinoaceae	*				
	Blasiaceae		*			
	Aneuraceae	*			*	
	Vandiemeniaceae					*
	Metzgeriaceae		*	*		
	Hymenophytaceae	*				
Treubiales	Treubiaceae	*				
Jungermanniales	Herbertaceae		*			
	Trichocoleaceae	*			*	
	Grolleaceae					*
	Trichotemnomaceae					*
	Antheliaceae	*				
	Vetaformaceae					*
	Lepicoleaceae					*
	Jungermanniaceae	*			*	
	Gymnomitriaceae	*				
	Scapaniaceae	*				
	Blepharidophyllaceae					*
	Delavayellaceae					*
	Geocalycaceae	*			*	
	Plagiochilaceae	*				
	Acrobolbaceae					*
	Arnelliaceae	*				
	Brevianthaceae					*
	Chonccoleaceae					*
	Schistochilaceae	*				
	Perssoniellaceae					*
	Balantiopsidaceae					*

Table 1. (*contd*)

Order	Family	Association with fungi				
		Regular or frequent		Occasional		No data
		+	−	+	−	
Jungermanniales (*contd*)	Gyrothyraceae					*
	Lepidoziaceae	*			*	
	Phycolepidoziaceae					*
	Calypogeiaceae	*			*	
	Cephaloziaceae	*				
	Cephaloziellaceae	*				
	Jackiellaceae	*				
	Adelanthaceae					*
	Ptilidiaceae	*				
	Mastigophoraceae					*
	Chaetophyllopsidaceae					*
	Lepidolaenaceae					*
	Jubulopsidaceae					*
	Neotrichocoleaceae					*
	Goebeliellaceae					*
	Porellaceae	*				
	Jubulaceae					*
	Lejeuneaceae		*			
	Radulaceae	*			*	
	Pleuroziaceae					*
Sphaerocarpales	Riellaceae					*
	Sphaerocarpaceae		*			
Monocleales	Monocleaceae	*				
Marchantiales	Cleveaceae		*	*		
	Aytoniaceae		*	*		
	Lunulariaceae	*			*	
	Conocephalaceae	*				
	Marchantiaceae	*			*	
	Monoseleniaceae					*
	Targioniaceae		*		*	
	Cyathodiaceae	*				
	Carrpaceae					*
	Corsiniaceae		*			
	Oxymitraceae		*			
	Ricciaceae		*	*		

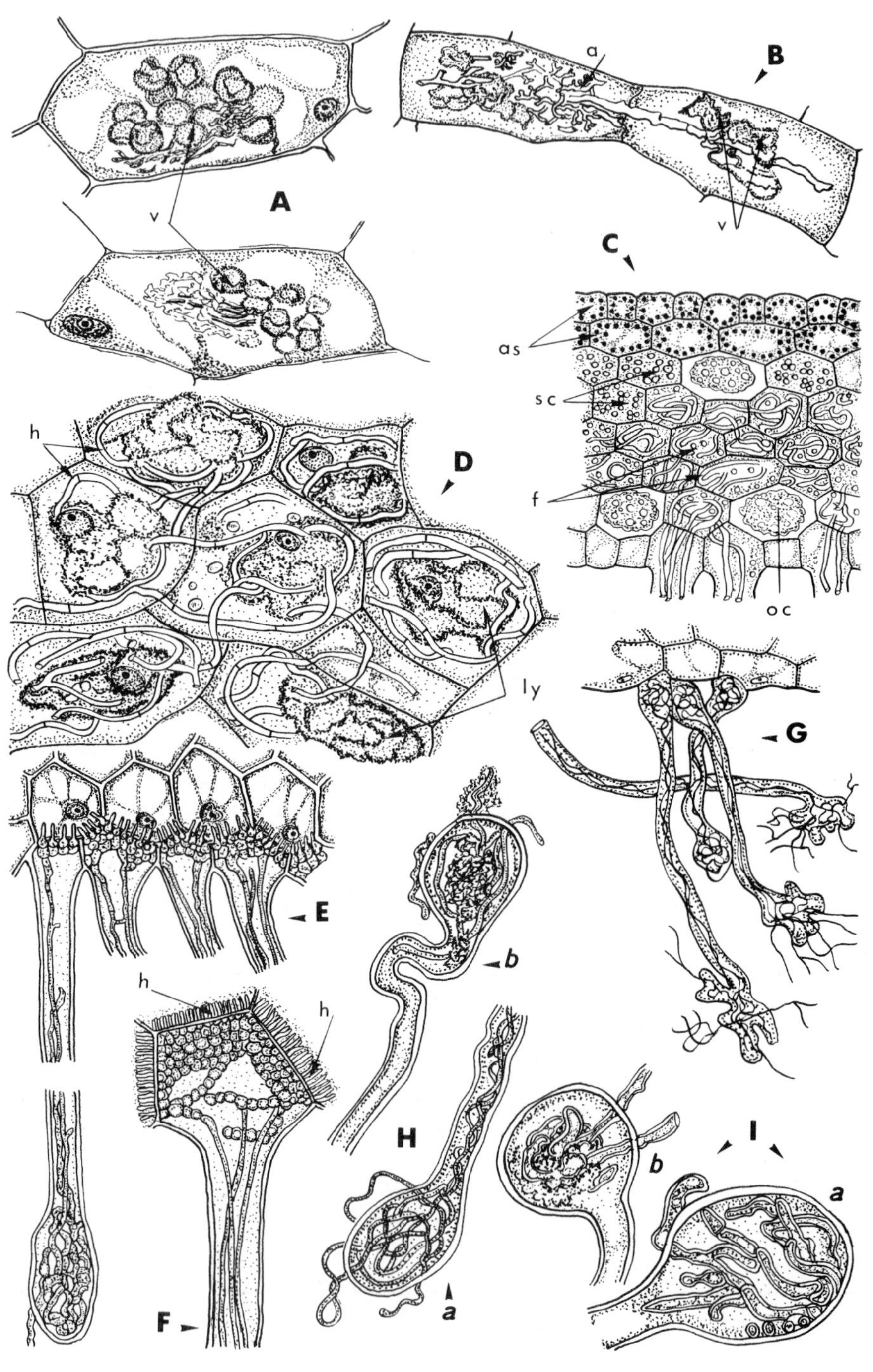
A
v
B
a
v
C
as
sc
f
oc
D
h
ly
E
F
G
H
a
b
I
a
b
h

worts, and even less about the cellular aspects of the interaction. Nevertheless, from the publications of Stahl (1949), and especially Pocock *et al.* (1984) and Pocock and Duckett (1985a,b), it is also clear that the very common and morphologically varied endophytes of the Hepaticae belong to different taxa of fungi as follows:

(a) dikaryotic basidiomycetes lacking clamp connections and forming intracellular pelotons ("tolypophagous"-type) in the stems of several families of the Jungermanniales and Aneuraceae (Metzgeriales);
(b) fungi (possibly Endogonaceae) forming thamniscophagous mycothalli in other Metzgerialian families and in the Marchantiales;
(c) fungi, the exact affinities of which are still unknown, in the rhizoids of several families of the Jungermanniales.

With regard to the endosymbiotic basidiomycetous fungi (already known from the root-cortex of the Orchidaceae), it has been suggested that the partner of *Cryptothallus mirabilis*, an achlorophyllous hepatic, is the ectomycorrhizal associate of the neighbouring birch and beech trees (Hadley, 1986).

With regard to thamniscophagous associations, although Pocock and Duckett (1984) hesitate to suggest it, I agree with Stahl (1949) that the fungi are chiefly (but not exclusively) species of the Endogonaceae. Aseptate hyphae, arbuscules, ptyosomes and true vesicles, like those of VA mycorrhizas in phanerogams, commonly occur in the hepatics. It must be noted, however, that Carré and Harrison (1961) demonstrated that a strain of *Pythium* reproduced the appearance of the vesicular-arbuscular endophyte in *Conocephalum conicum*.

Discomycetous fungi have been suspected to be involved with the Jungermanniales and the Marchantiales; for example, *Mollisia jungermanniae* with

---

Figure 2. A, *Treubia insignis*, digested vesicles (v). Thamniscophysalidophagous mycothallus (after Stahl, 1949). **B**, *Symphyogyne* sp., arbuscule (a) and digested vesicles (v). Thamniscophysalidophagous mycothallus (after Stahl, 1949). **C**, *Monoclea forsteri*, transverse section of mycothallus with coils of hyphae (f); oc, oil-containing cell; sc, starch-grain cells; as, assimilating cells (after Cavers, 1904). **D**, *Cryptothallus mirabilis*, coiled hyphae (h) and residual material after lysis (ly). Tolypophagous mycothallus. **E**, *Calypogeia trichomanes*, transverse section of basal part of stem and rhizoids with fungi bearing haustoria in the basal part and coiled hyphae in the apical part of a rhizoid. Haplohaustorial mycothallus (after Némec, 1899). **F**, *Lophozia lycopodioides*, base of a rhizoid harbouring endophytic fungus; h, haustoria. Haplohaustorial mycothallus (after Stahl, 1949). **G**, *Cephalozia connivens*, infected rhizoids with swollen forked apices. **H**, *Microlepidozia silvatica*, apices of rhizoids containing intact (a), and lysed (b) hyphae. **I**, *Telaranea setacea*, swollen apices of rhizoids containing intact (a) and lysed (b) hyphae.

Table 2. Some examples of morphological types of mycothallic associations

Thamniscophagous mycothalli	
Androcryphia confluens	*Lunularia cruciata*
Conocephalum conicum	*Marchantia calcarea*
Dumortiera irrigua	*Marchantia geminata*
Fossombronia angulosa	*Möerckia blyttii*
Fossombronia incurva	*Monoclea forsteri*
Haplomitrium gibbsiae	*Pallavicinia sciphoides*
Lophocolea leucophylla	*Pellia epiphylla*
Thamniscophysalidophagous mycothalli	
Haplomitrium hookeri	*Symphyogyne* sp.
Pallavicinia sp.	*Treubia insignis*
Tolypophagous mycothalli	
Alicularia scalaris	*Cryptothallus mirabilis*
Aneura maxima	*Riccardia chamaedryfolia*
Aneura pinguis	*Riccardia incurvata*
Haplohaustorial mycothalli	
Anthelia juratzkiana	*Haplozia sphaerocarpa*
Calypogeia trichomanes	*Lophozia barbata*
Gymnomitrium concinnatum	*Lophozia lycopodioides*
Mycothalli with infected apically swollen rhizoids	
Cephalozia connivens	*Psiloclada cladestina*
Cephalozia lammersiana	*Telaranea setacea*
Mylia anomala	*Telaranea silvatica*
Nardia scalaris	*Zoopsis liukiuensis*

Calypogeia trichomanes (Nemec, 1899); *Mniaecia jungermanniae* with *Alicularia scalaris* (Killian, 1926); *Gloeopeziza rehmii* with *Jungermannia trichophylla*; *G. crozalsi* with *Calypogeia ericetorum*; and *Humaria nicolai* with *Lunularia cruciata* (Nicolas, 1932). These are probably examples of casual or weakly parasitic associations comparable to the more recently described (Döbbeler, 1980) *Epibryon endocarpum* on *Plagiochila asplenioides*, *Chladochytrium aneurae* and *Aneura* sp., or *Olpidiopsis ricciae* and *Riccia* sp.

THE SIGNIFICANCE OF FUNGAL SYMBIOSIS

It is reasonable to postulate that mutualistic fungal symbionts have facilitated the colonization of inhospitable biotopes by some liverworts, as already implied for Marchantiaceae, but can this claim be substantiated? Considering only the Marchantiales and taxa better suited to more or less wet sites

(Metzgeriales, Jungermanniales, Monocleales, Sphaerocarpales), one could almost agree. However, the environment does not act alone. Of the three families inhabiting seasonally arid areas, the Fossombroniaceae are almost constantly mycotrophic, but mycothalli do not occur in the Oxymitraceae and appear to be very scarce in the Ricciaceae. On the other hand, liverworts growing in strongly oligotrophic bogs are commonly mycotrophic with "the fungi having an important absorptive role" (Pocock *et al.*, 1984).

Although data are still lacking for most species of Hepaticeae, some provisional conclusions can be drawn:

(a) Some hepatics are consistently free of fungi. Such species usually represent more advanced taxa within their respective phyla (e.g., *Blasia pusilla*, Stahl, 1949; Pocock and Duckett, 1984; *Clevea rousseliana*, Golenkin, 1902; *Metzgeria furcata*).
(b) In other species, the association with fungi appears to be either favourable or injurious, depending on the growth conditions (e.g., *Conocephalum conicum*, Peklo, 1903; *Möerckia blyttii*, Stahl, 1949).
(c) Physiological studies utilizing radioisotopes, coupled with analysis and synthesis of mycothalli (cf. Stahl, 1949), are needed to provide less equivocal data.
(d) Nevertheless, it is certain that fungal parasitism exists in the Hepaticeae, as in *Bucegia romanica* (Eftimiu, 1933); *Anthelia juratzkiana* (Stahl, 1949), or *Plagiochila asplenioides* (Döbbeler, 1980). It is equally certain that some fungi are "neutral" commensalists or form "casual" unions (which explains inconsistent results for different species) as in *Aneura palmata*, *A. sinuata* (Stahl, 1949) and *Marchantia palaeacea* (Golenkin, 1902; Stahl, 1949).
(e) Yet, it must also be asserted that cases of mutualistic symbiosis exist, undoubtedly benefiting the host, and surely, in exchange, the fungal associate (as in *Marchantia nepalensis*; see below).

Cavers (1903a) emphasized the permanence and the intimacy of integration in certain mycothalli of *Concocephalum conicum*: "in some cases there can be no doubt that the association of the fungus with the liverwort is of a symbiotic nature". *Marchantia nepalensis* cannot reproduce sexually without its fungal symbiont (Chaudhuri and Rajaram, 1926), and it may prove instructive to re-examine in this connection the behaviour of, for example, *Lunularia cruciata* or *Pellia epiphylla*.

Undoubtedly in VA mycorrhizae the lysis of hyphae, arbuscules or vesicles provides the host with various valuable products of fungal metabolism. As suggested by Cavers (1903b) for *Conocephalum conicum* and Stahl (1949) for species of *Marchantia*, *Pellia* and *Riccardia*, the effect of fungi upon growth or fleshiness of thalli is beyond doubt. However, in the absence of data on the specificity of associations, integration (i.e., the appearance of novel chemical characters), pathways of reciprocal nutrient exchange, etc., it is difficult to quantify the effect of coevolution.

Almost constant infection leads to the formation of (modest but unquestionable) tubers in *Fossombronia*. Tubers appear to be better developed in *Sewardiella*, especially *S. tuberifera* (Chalaud, 1932). The latter was described by Chalaud as "le terme le plus parfait de l'adaptation d'un gamétophyte pérennant à la vie en commun avec un Champignon."

Cryptothallus mirabilis, with colourless and almost rhizoidless thalli, which from the biological point of view is very similar to the phanerogams *Neottia nidus-avis* or *Monotropa hypopitys*, is a liverwort "dependent on its mycorrhizal fungus for the provision of organic carbon" (Hadley, 1986). This wholly achlorophyllous holosaprophyte can be biologically compared with underground pteridophytes (Boullard, 1979). It is dependent, in a way, upon its fungal endophyte. Similar trophic relationships can be found, scattered through the Hepaticae. The relationship is clearly antagonistic but it does not preclude coevolution, even if the hepatic appears to be the sole beneficiary to the point of developing obligate dependence on its fungal associate.

In *Haplomitrium*, *Fossombronia*, *Cryptothallus* and several other hepatics, when an external factor (here fungal infection) exerts an uninterrupted evolutionary pressure through successive generations, the trait is not thus far hereditary but nevertheless enhances survival of the host, and the resultant morphosis acquires the value of "permanent specific features". Such "gained" characters appear as being "fixed" (Magrou, 1948, p. 363). Thus, by influencing morphogenetic expression, fungal symbiosis profoundly affects the evolutionary fitness of the species. I claim, therefore, that the appearance of an obligately symbiotic liverwort may be independent of the usually omnipotent genotypic constitution of the plant. Biological duality of mycothallic species offers new evolutive potentialities, and provides a glimpse of a fascinating aspect of coevolution.

If the common ancestor of bryophytes and pteridophytes had an isomorphic lifecycle (Schuster, 1984, p. 913), and if constant mycotrophy in the Hepaticae is primitive, a parallel can be drawn between the gametophytes of Hepaticeae and those of Schizaeaceae, Lycopodiaceae, *Psilotum* and *Rhynia*, one of the main peculiarities of which is constant mycotrophy.

In conclusion, though mycothallic associations are somewhat plastic *vis-à-vis* environmental conditions, symbiosis in the Hepaticae is very old. Because "it is difficult to see any clear relationship between the presence of fungal endophytes and ecology" (Pocock and Duckett, 1985c), one is forced to admit that fungal symbiosis in the Hepaticae is very old, and to recognize that coevolution with fungi must be a significant aspect of their evolution and biology.

REFERENCES

Auret, T. B. (1930). Observations on the reproduction and fungal endophytism of *Lunularia cruciata* (L.) Domortier. *Trans. Br. mycol. Soc.* **15,** 163–176.

Bernard, N. (1910) ["1909"]. L'évolution dans la symbiose. *Ann. Sci. nat., Bot.* 1–196.

Bolleter, E. (1905). *Fegatella conica* (L.) Corda. Eine morphologisch-physiologische Monographie. *Beih. Bot. Zbl.* **18,** 327–408.

Boullard, B. (1960). Licht und Mykorrhiza. In "Mykorrhiza, Internationales Mykorrhiza Symposium" (C. Fischer, ed.), pp. 315–329. Jéna.

Boullard, B. (1979). Considérations sur la symbiose fongique chez les Ptéridophytes. *Syllogeus* **19,** 1–59.

Burgeff, H. (1938). Mycorrhiza. In "Manual of Pteridology" (F. Verdoorn, ed.), pp. 159–191. Martinus Nijhoff, The Hague.

Campbell, E. O., Markham, K. R., Moore, N. A., Porter, L. J., and Wallace, J. W. (1979). Taxonomic and phylogenetic implication of comparative flavonoid chemistry of species in the family Marchantiaceae. *J. Hattori Bot. Lab.* **45,** 185–199.

Carré, C. G. and Harrison, R. W. (1961). Studies on vesicular–arbuscular endophytes. III. *Trans. Brit. Mycol. Soc.* **44,** 565–572.

Cavers, F. (1903a). On saprophytism and mycorrhiza in Hepaticae. *New Phytol.* **2,** 20–35.

Cavers, F. (1903b). On the structure and biology of *Fegatella conica. Ann. Bot.* **18,** 87–120.

Cavers, F. (1904). On the structure and development of *Monoclea Forsteri* Hooker. *Revue bryol.* **31**(4), 69–80.

Cavers, F. (1911). The inter-relationships of the Bryophyta. *New Phytol.* **10,** 1–46.

Chadefaud, M. and Emberger, L. (1960). "Traité de Botanique Systématique." Vol. I. "Les Végétaux non-vasculaires." Masson, Paris.

Chalaud, G. (1932). Mycorhizes et tubérisation chez *Sewardiella tuberifera* Kashyap. *Annls. bryol.* **5,** 1–16.

Chaudhuri, H. and Rajaram, M. (1926). Ein Fall von warscheinlicher Symbiose eines Pilze mit *Marchantia nepalensis. Flora, Jena* **120,** 176–178.

Crandall-Stotler, B. (1984). Musci, hepatics and anthocerotes—an essay on analogues. In "New Manual of Bryology" (R. M. Schuster, ed.), vol. 2, pp. 1093–1129. Hattori Botanical Laboratory, Nichinan.

Döbbeler, P. (1980). *Epibryon endocarpum* sp. nov. (Dothideales), ein hepaticoler Ascomycet mit intrazellulären Fruchtkörpen. *Z. Mykol.* **46,** 209–216.

Edwards, D. F. (1986). *Aglaophyton major*, a non-vascular land-plant from the Devonian Rhynie Chert. *Bot. J. Linn. Soc.* **93,** 173–204.

Eftimiu, P. (1933). Sur la présence d'un champignon chez *Bucegia romanica* Radiam. *C. r. hebdom. Séanc. Acad. Sci., Paris* **196,** 957–959.

Evans, A. W. (1939). The classification of the hepatics. *Bot. Rev.* **5,** 49–96.

Fulford, M. (1965). Evolutionary trends and convergence in the Hepaticae. *Bryologist* **68**, 1–31.

Golenkin, M. (1902). Die Mycorrhiza-ähnlichen Bildungen der Marchantiaceen. *Flora, Jena* **90,** 209–220.

Hadley, G. (1986). Mycorrhizas of heterotrophic plants. In "Physiological and Genetical aspects of Mycorrhizae" (V. Gianinazzi-Pearson and S. Gianinazzi, eds.), pp. 815–819. INRA, Paris.

Kashyap, S. R. (1919). The relationship of liverworts, especially in the light of some recently discovered Himalayan forms. *Proc. Asiatic Soc. Bengal*, n.s., **15,** clii–clxvi.

Killian, C. (1926). Nouvelles observations sur la symbiose du *Mniaecia jungermanniae* avec *Alicularia scalaris. Bull. Ass. Philomat. Alsace-Lorraine* **7,** 146–150.

Lowry, B., Lee, D., and Hébant, C. (1980). The origin of land plants: a new look at an old problem. *Taxon* **29,** 2–3, 183–197.

Magrou, J. (1948). La symbiose, facteur d'évolution dans le règne végétal. *Rev. Quest. Scient.*, 20 July 1948, 340–371.

Mehra, P. N. (1968). Phyletic evolution in the Hepaticae. *Phytomorphology* **17,** 47–58.

Mehra, P. N. (1969). Evolutionary trends in Hepaticae with particular reference to the Marchantiales. *Phytomorphology* **19,** 203–218.

Némec, B. (1899). Die Mykorrhiza einiger Lebermoose. *Ber. dt bot. Ges.* **17,** 311–317.

Nicolas, G. (1932). Associations des Bryophytes avec d'autres organismes. In "Manual of Bryology" (F. Verdoorn, ed), pp. 109–124. Martinus Nijhoff, The Hague.

Peklo, J. (1903). Kotacze mykorhizy u muscinei. *Rozpr. Abhandl. böhm. Akad. Prague* **12,** 32–40.

Peyronel, B. (1939). Luce, humus e micorrizia. *Atti r. Accad. Sci. Torino* **75,** 1–13.

Pocock, K. and Duckett, J. G. (1984). A comparative ultrastructural analysis of the fungal endophyte in *Cryptothallus mirabilis* Malm. and other British thalloid hepatics. *J. Bryol.* **13,** 227–233.

Pocock, K. and Duckett, J. G. (1985). The alternative mycorrhizas: fungi and hepatics. *Bull. Br. bryol. Soc.* **45,** 1 page (unnumbered).

Pocock, K. and Duckett, J. G. (1985b). Symbiotic relationships between liverworts and endophytic fungi. In "Abstracts." IIIrd International Congress of Systematics and Evolutionary Biology, Brighton, 4–10 July, p. 149. International Congress of Systematic and Evolutionary Biology, Brighton.

Pocock, K. and Duckett, J. G. (1985c). Fungi in hepatics. *Bryological Times* **31,** 2–3.

Pocock, K., Duckett, J. G., Grolle, R., Mohamed, M. H. H., and Pang, W. C. (1984). Branched and swollen rhizoids in hepatics from montane rain forest in Peninsular Malaya. *J. Bryol.* **13,** 241–246.

Schuster, R. M. (1979). The phylogeny of the Hepaticae. In "Bryophyte Systematics" (Clarke G. C. S. and Duckett J. G., eds), pp. 41–82. Academic Press, London.

Schuster, R. M. (1983). "New Manual of Bryology." Vol. 1. Hattori Botanical Laboratory, Nichinan.

Schuster, R. M. (1984). "New Manual of Bryology." Vol. 2. Hattori Botanical Laboratory, Nichinan.

Stahl, M. (1949). Die Mycorrhiza der Lebermoose mit besonderer Berücksichtigung der thallosen Formen. *Planta* **37,** 103–148.

6 Coevolution of Fungi with Algae and Cyanobacteria in Lichen Symbioses

D. L. HAWKSWORTH

CAB International Mycological Institute, Kew, Surrey, UK

Abstract

A review of the range of mutualistic mycobiont–photobiont symbioses, and the systematic position of the bionts occurring in lichen symbioses, is presented. Coevolved structures and products are recognized: the vegetative morphology of lichen thalli (i.e., lichenized stromata); the production of specialized dual propagules ensuring the simultaneous dispersal of both bionts; the contact sites between the bionts as examined at the ultrastructural level; physiological and primary metabolic adaptation to otherwise hostile environments; and novel secondary metabolites, some reportedly requiring both bionts for their biosynthesis. Coevolution is also seen in the development of asexual strategies for reproduction and dispersal, niche exploitation, longevity, and in relationships with certain invertebrates and other organisms. The evolutionary significance of lichen symbioses is discussed with an emphasis on ascomycetes, the ancestors of which are considered to have been rock- and soil-dwelling lichen symbionts associated with cyanobacteria.

INTRODUCTION

Lichenization is a major method of nutrition in the fungi, indeed about one-fifth of all fungi, that is around 13 500 species, enter into lichen associations. It is a method of nutrition seen in more fungi than those that form mycorrhizas, yet its evolutionary significance is rarely considered. Discussions of how such relationships have evolved and of their significance for the bionts involved have been generally avoided by recent authors, with a few notable exceptions (e.g., Ahmadjian, 1970, 1987; Cain, 1972).

The evolution of a lifestyle that can dominate huge areas in the boreal zones, and also enable fungi to persist in some of the hottest deserts of the world, in the highest mountain ranges, in tropical rain forests, and even in-

COEVOLUTION OF FUNGI
WITH PLANTS AND ANIMALS
ISBN 0-12-557365-0

thallus and then being incorporated into discrete cephalodia or lobules. A more bizarre case is the lichenicolous lichen *Buellia pulverulenta* on the lichen *Physconia distorta*, where the *Buellia* has a discrete photobiont embodied within the thallus of the host *Physconia* (Hafellner and Poelt, 1980). Lichenicolous lichens have been much overlooked but are becoming increasingly recognized (e.g., Poelt, 1985; Steiner and Poelt, 1987); some always remain lichenicolous but others eventually develop totally independent thalli leaving no trace of their original hosts.

THE BIONTS IN LICHEN SYMBIOSES

Mycobionts in lichens

Some 525 genera and 13 500 species of lichen-forming fungi are currently recognized (Hawksworth *et al.*, 1983); the names given to lichens strictly refer to the mycobiont (Hawksworth, 1978). These fungi are almost exclusively found in lichen associations and do not have independent lifestyles as saprobes or parasites; exceptions are seen only in facultatively lichenized species and some initially lichenicolous species. The mycobionts mainly belong to the Ascomycotina, almost half the class being lichenized. Sixteen of the 46 orders include lichenized representatives (Hawksworth, 1985), but only five of these orders are exclusively lichenized. This emphasizes the significance of this nutritional type in the evolutionary biology of the ascomycetes as a whole (Eriksson, 1981; Dick and Hawksworth, 1985); this aspect is considered further below.

Lichenization is scattered throughout other groups of fungi (Hawksworth, 1988b), also including the basidiomycetes, for example *Dictyonema* and *Multiclavula* species; conidial fungi (see Vobis and Hawksworth, 1981), as in the hyphomycetes *Blarneya hibernica* (which has characteristic "lichen" secondary metabolites), and *Cheiromycina flabelliformis* (Hawksworth and Poelt, 1986); and rarely the protoctistan mastigomycetes, with *Geosiphon pyriforme* (Mollenhauer, 1970), and perhaps *Lagenidium hyphinicola* (Moser-Rohrhofer, 1975), whose biology remains uncertain. Groups other than the Ascomycotina are rarely studied by lichenologists, however, and many lichenized conidial fungi, both Coelomycetes and Hyphomycetes, remain to be formally described, even from Europe. Lichenized conidial fungi have their affinities with the Ascomycotina and clearly represent anamorphs of either extinct or yet to be recognized teleomorphs.

Photobionts in lichens

In contrast to the number of mycobiont genera involved in lichen associ-

Table 1. Photobiont genera reported as serving as inhabitants in lichen associations

	Order	Genera
Cyanobacteria	Chamoesiphonales (incl. Pleurocapsales)	*Hyella*
	Chroococcales[a]	*Aphanocapsa*, *Chroococcus*, *Chroococcidiopsis*, *Gloeocapsa*
	Hormogonales (incl. Stigonematales)	*Anabaena*, *Calothrix*, *Dichothrix*, *Fischerella*, *Hyphomorpha*, *Nostoc*, *Scytonema*, *Stigonema*
Chlorophyta[c]	Chaetophorales	*Coccobotrys*, *Desmococcus* (syn. *Protococcus*), *Dilabifilium*[b], *Leptosira*, *Pseudopleurococcus*
	Chloroccocales	*Asterochloris*, *Chlorella*, *Chlorococum*, *Coccomyxa*, *Dictyochloropsis*, *Elliptochloris*, *Myrmecia*, *Pseudochlorella* (incl. *Chlorellopsis* sensu Zeitler), *Trebouxia* (incl. *Pseudotrebouxia*), *Trochiscia*
	Chlorosarcinales	*Chlorosarcina*, *Friedmannia*
	Tetrasporales	*Gloeocystis*
	Trentepohliales	*Cephaleuros*, *Phycopeltis*, *Physolinum*, *Trentepohlia*
	Ulotrichales	*Stichococcus*
Phaeophyta	Ectocarpales	*Petroderma*
Xanthophyta	Mischococcales	*Botrydiopsis*
	Trichonematales	*Heterococcus*

[a]Reported *Anacystis* photobionts will probably prove to belong to *Aphanocapsa* or *Gloeocapsa* (B. A. Whitton, *in litt.*).
[b]Perhaps a synonym of *Pseudoendoclonium* (Bourrelly, 1973).
[c]*Hyalococcus* is regarded as incompletely described and of uncertain identity (Bourrelly, 1973).

ations, the number of photobiont genera is minute. Only about 39 (Table 1) are currently accepted and some of those are of questionable taxonomic value. Such a disparity in numbers between exhabitants and inhabitants is a general characteristic of mutualistic symbioses (Law, 1985). Interestingly, all the cyanobacterial genera involved occur free in nature, as do all but two or three of the green-algal ones (that number might actually be only one as *Dilabifilium* and *Hyalococcus* are of doubtful status (Bourelly, 1973)). Two of the commonest green photobiont genera, *Myrmecia* and *Trebouxia*, are rarely found free-living (Tschermak-Woess, 1978), and such colonies are probably generally derived from lichen dual propagules (see below). Ahmadjian (1962) found that some *Trebouxia* isolates only remained healthy in culture when enveloped with mycobiont hyphae, and doubted their independent existence (Ahmadjian, 1987). The consensus is that it is unlikely that they can sustain a free-living existence indefinitely (Smith and Douglas, 1987). Generic concepts in both non-marine green algae and cyanobacteria are currently confused, and in many cases uncertainty surrounds the placement of photobionts from lichen thalli. However, *Trebouxia* could have been derived from a free-living *Pleurastrum* according to the pyrenoid structure (Molnar *et al.*, 1975).

The confusion increases at the species level. Species concepts in the photobionts of lichens are extremely uncertain, so much so that specificity at this level cannot be discussed in any meaningful way at this time. As noted by Ahmadjian (1982), "many species of lichen algae . . . must be considered suspect because they were not based on the now accepted standardized methods of culture". This is a major area open for original research, but it must be stressed that pure culture studies are needed in order to determine the true morphology and both sexual and asexual reproductive stages of the photobionts concerned. Inside lichens, filamentous algae tend to become unicellular and reproduction is largely asexual and by aplanospores. Zoospores have been found within thalli in a few cases (Slocum *et al.*, 1980), but no firm evidence for sexual reproduction within lichen thalli has been obtained. This is in accordance with the general theory of sexuality in inhabitants proposed by Law and Lewis (1983).

However, apart from the cyanobiont/phycobiont switches noted above, there is now evidence that some mycobiont species can form identical thalli with different genera of green alga; for example, *Chaenotheca carthusiae* is lichenized with *Trebouxia* in Costa Rica and not with *Stichococcus* as in the Northern Hemisphere (Tibell, 1982). It is also becoming clear that single mycobiont species can form morphologically identical thalli with different species of the same photobiont genus. This situation could even be the rule rather than the exception. Examples are *Coenogonium linkii* with different *Trentepohlia* species (Uyenco, 1965); *Xanthoria parietina* with different *Trebouxia* species (Piattelli and de Nicola, 1968; Ott, 1987a); *Diploschistes muscorum* with two *Trebouxia* species (Friedl, 1987); and *Parmelia acetabulum*, which may have one of at least three species of *Trebouxia* (T. Friedl, personal communication). On the other hand, the same photobiont species can associate with different mycobionts, forming morphologically different thalli, as with *Cephaleuros virescens* in various foliicolous lichens belonging to the Strigulaceae (Santesson, 1952), and *Trebouxia excentrica* in at least five genera and seven lichen species (Hildreth and Ahmadjian, 1981).

COEVOLVED STRUCTURES AND PRODUCTS

In view of the plasticity and varying degrees of specificity described above, to what extent can the partners involved in lichen associations be considered to have coevolved? If coevolution is interpreted in the sense of genetic changes in both partners in response to each other, resulting in increased overall evolutionary fitness, it must be stressed that hard evidence for such changes is not available for lichen symbioses. When a genetic change in one partner occurred that increased the fitness of the symbiosis as a whole, it would tend to be selected for and fixed in the population. Over the time span in which lichen taxa have evolved (see p. 143), it is not unreasonable to speculate that such events have occurred from time to time. Indeed, there are indications that coevolution has contributed to the variety and success of the lichen life-style in the vegetative morphology, dual propagules, biont contacts, physiology and primary metabolism, and secondary metabolism.

Vegetative morphology

The most obvious innovation in lichen associations is the thallus itself. The variety of thallus forms can be interpreted as coevolved methods of displaying the inhabiting photobionts to ensure optimal exposure to light and so maximization of photosynthesis to the benefit of both partners. Indeed, the presence of a photobiont can stimulate the production of a cortex in thalli turned upside-down (Ahmadjian, 1987).

When grown in pure culture on agar, lichen mycobionts almost exclusively produce circular, amorphous, slow-growing colonies similar to those encountered in parasitic and saprobic fungi that are unstratified in section, and not differentiated into tissue types (Ahmadjian, 1967). It is only in the presence of a suitable photobiont that development of a thallus recalling that encountered in nature is initiated. This has now been demonstrated in culture for a variety of species, including *Cladonia cristatella* (Ahmadjian *et al.*, 1980), *Endocarpon pusillum* (Ahmadjian and Heikkilä, 1970), *Heppia echinulata* (Marton and Galun, 1976), *Rhizoplaca chrysoleuca* (Culberson and Ahmadjian, 1980), *Usnea strigosa* (Ahmadjian and Jacobs, 1985), *Verrucaria macrostoma* (Stocker-Wörgötter and Türk, 1987), and *Xanthoria parietina* (Bubrick and Galun, 1986).

There is now some evidence that this applies in at least some cases in the field. Ott (1987b) found that *Xanthoria* mycobionts could exist in loose crustose associations with a variety of algae until an appropriate *Trebouxia* was available with which foliose lobes could be formed; only then was thallus differentiation initiated. While the effect on thallus form of different species from a single algal genus, or even in some cases of different genera of the same group of algae, consequently appears to be of little significance, an appropriate photobiont is vital to development. However, clear indications of the relevance of the photobiont to thallus morphology are seen in morphotypes produced by cyanobacteria rather than green algae with the same mycobiont (see p. 127). In the case of the *Solorina spongiosa* group, the type of thallus found depends on whether an appropriate alga or cyanobacterium is encountered first (Jahns, 1988).

The nature of the recognition mechanisms between the bionts leading to thallus formation is almost entirely unknown. However, ultrastructural studies have demonstrated that surface layers of the walls of both mycobionts and photobionts from the same lichen have tesellated patterns that mesh together when contacts are made. These develop on the photobiont cell surfaces before contacts are made and seem to be a regular feature of adult photobiont cells; they have been demonstrated in thalli of *Chaenotheca*, *Cladonia*, *Hypogymnia*, *Lobaria* and *Parmelia* (Honegger, 1984). Such surface structures may be involved in immunological recognition mechanisms suggested by antiserum investigations to occur between bionts (Bubrick *et al*., 1981). However, it is uncertain whether lectin-mediated recognition is important in the contact phases of the establishment of lichen symbioses (Lallement and Savoye, 1985; Smith and Douglas, 1987).

The differentiated thalli can be regarded as secondary or tertiary structures (Lallement *et al*., 1987). Their production has been compared by Moreau (1956) to that of gall formation, as novel structures resulting from a photobiont's presence. As pointed out by James and Henssen (1976), this parallel should not be dismissed; indeed, experimental studies on gall formation in other groups might yield insights into the mechanisms of thallus formation in lichenized fungi. The mechanisms controlling the development of distinctive thallus shapes seen only in symbioses are obscure, but the studies of Greenhalgh and co-workers (see Greenhalgh and Whitfield, 1987) indicate that the hyphae of the mycobiont grow between and separate aplanospores of the photobiont that then grow to their usual size, the process being repeated. It is uncertain how synchrony is achieved and whether the mycobiont growth to provide "space" for new photobiont cells is limiting or the rate of division and carbohydrate products of the latter. In the developed thallus, however, while the rate of division of algal cells decreases, their mean size increases and the number of empty dead cells declines in *Bryoria fuscescens* (Greenhalgh and Whitfield, 1987). There are indications that the rate of growth of the photobiont cells is reduced by some probably mycobiont-derived inhibitory product not yet identified with certainty (Honegger, 1987).

It is of interest that some of the more elaborate thallus shapes produced in some lichens have parallels in the stromata of certain non-lichenized fungi never found in association with algae. For example, radiating convex lobes occur in *Hypocreopsis* (Hypocreales), branched fruticose structures in certain groups of *Xylaria* (Xylariales), and thread-like or strap-like filaments in *Thamnomyces* (Xylariales). The possibility that such structures are homologous with lichen thalli should not be excluded. Although these are now formed in the absence of any inhabiting photobiont, they could be interpreted as relics from long-extinct lichenized ancestors that produced such coevolved structures only in the presence of a photobiont.

Changes in morphology within the lichen thalli are not confined to the mycobiont, as noted above; even the size and shape of the algal cells can vary depending on the part of the lichen in which they occur in *Staurothele caesia* (Ahmadjian and Heikkilä, 1970). The long chains of cyanobacterial cells seen in nature in *Nostoc* may also become disassociated into shorter ones (Degelius, 1954); the frequency of nitrogen-fixing heterocysts increases where these are confined to discrete cephalodia (Kershaw, 1985); and the life cycle of the *Nostoc* within the thallus is restricted (Lallement *et al.*, 1987). These differences, together with suppressed sexuality (see p. 130), hinder the determination of lichen photobionts. Ultrastructural changes, for example in wall thickness, thylakoid patterns in the pyrenoids, and numbers of pyrenoglobuli, also occur in the lichenized state (Ahmadjian and Jacobs, 1983).

Dual propagules

Many lichens have evolved specialized dual propagules to affect the dispersal of both mycobiont hyphae and compatible photobiont cells as a single unit. These dual propagules take many forms, such as: groups of algal cells enclosed in wefts of fungal hyphae ("soredia") formed irregularly or in discrete specialized structures ("soralia"); corticate lobate or scale-like ("phyllidia"), coralloid or subglobose, outgrowths ("isidia"); localized patches of cortical and photobiont tissues splitting away from the thallus surface ("schizidia"); basally constricted specialized branches breaking away ("spinules", "fibrils"); chains of nodular hyphae with intermixed algal cells in certain foliicolous lichens ("hyphophores"); and portions of thalli being cut off the main thallus ("fragmentation"). An exhaustive survey of these structures falls outside the scope of this contribution, but examples of some of these types are shown diagramatically in Figure 1. Further information on these structures may be found in Hawksworth and Hill (1984), and Jahns (1974); their major evolutionary significance is considered separately below.

Biont contacts

The type of contact between the bionts within lichen thalli can be of several types. Major progress in our understanding of the contact points at the ultra-

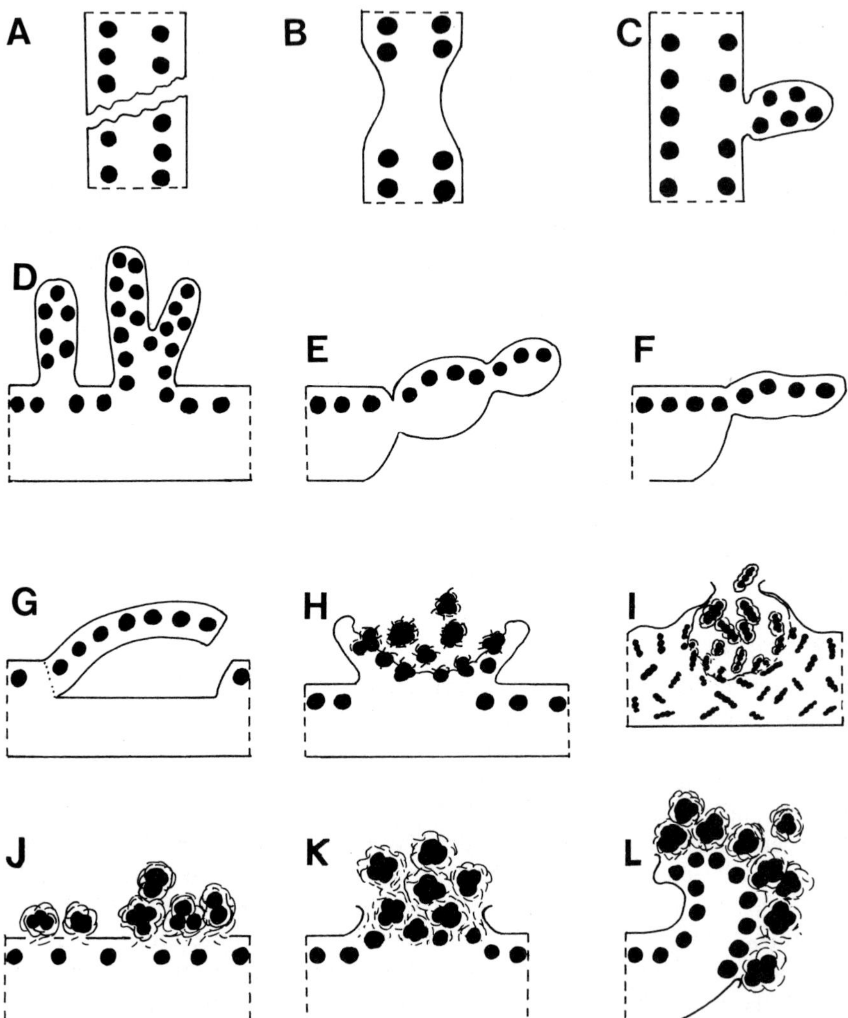

Figure 1. Diagrammatic representation of the various types of dual propagules produced by lichens. **A**, Thallus fragmentation; **B**, fragmentation region; **C**, lateral spinules; **D**, isidia; **E**, blastidia; **F**, phyllidium; **G**, schizidium; **H**, goniocyst; **I**, hormocyst; **J**, soredia formed from eroded surface; **K**, soredia formed in soralium; **L**, soredia formed from recurved lower cortex (labriform). Reproduced from Hawksworth and Hill (1984).

structural level have been made by Honegger (1986a,b). The nature of the contact site is much more specialized in some groups of lichens than in others. In orders and families which are believed on other grounds to be more recently lichenized, wall-to-wall apposition or intracellular haustoria penetrating into the photobiont cells can be found, whereas in more anciently lichenized groups, such as the *Trebouxia*-containing Parmeliaceae, highly specialized interparietal haustoria are developed (Figure 2). Haustoria of type 3 particularly will maximize the transfer of metabolic products between the bionts while minimizing the extent of any damage to the photobiont cells. A common gelatinous sheath forms over the whole contact area (Figure 3), enveloping the photobiont cells and also bearing crystals of complex secondary metabolites often peculiar to the symbiosis and not formed by either partner alone (see p. 139).

The photobiont cell walls of *Coccomyxa* and *Myrmecia* are three-layered; the outermost layer contains a highly resistant sporopollenin-like material that may afford some sort of defence against the penetration of fungal hyphae (Honegger and Brunner, 1981). Similar results have also now been obtained for phytobionts from the genera *Elliptochloris*, *Pseudochlorella*, *Trebouxia* and *Trentepohlia*, indicating that this situation is widespread (Brunner and Honegger, 1985). Walls of both partners may also be tesellated to facilitate recognition and adhesion (see p. 132).

The relationship between complex interparietal haustorium type and resistant photobiont cell walls is most appropriately regarded as coevolved over an extended time-scale. The situation is different from the gene-for-gene "arms race" coevolution between pathogens and hosts (Parlevliet, 1986; Hijwegen, this volume, Chapter 3), in that there are mutual advantages (see p. 142). However, mutualistic symbioses are recognized in other associations as generally being derived from parasitic ones (Thompson, 1986; Smith and Douglas, 1987), a shift in the interaction placing the mutualism at an advantage, especially in nutrient-poor environments (Lewis, 1973). The extent to which coevolution has been involved in determining the type of biont contact does, however, clearly vary from group to group and appears to be related to the antiquity of particular lichen symbioses.

Attention is also drawn here to the extensive intracellular haustoria formed by species of the basidiomycete genus *Dictyonema* within *Scytonema* filaments; the plasmalemma of the cyanobacterium invaginates along the haustorium and both bionts remain healthy (Slocum, 1980). A carefully balanced coevolved system can again be suspected here.

Physiology and primary metabolism

Evidence for physiological coevolution in lichens can be searched for at a

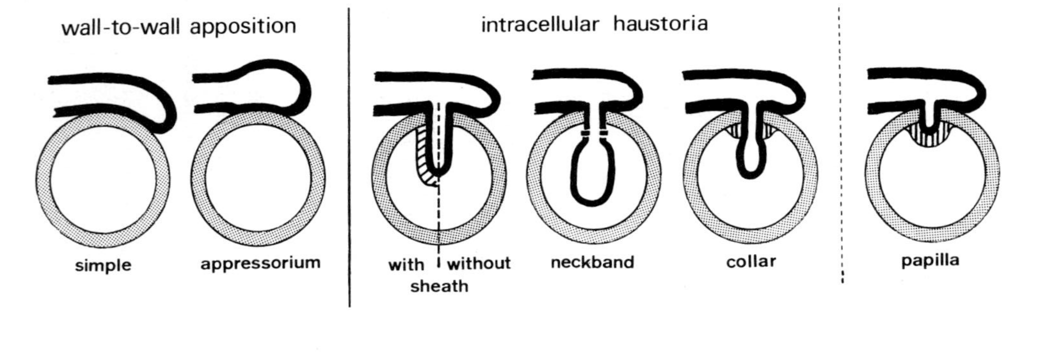

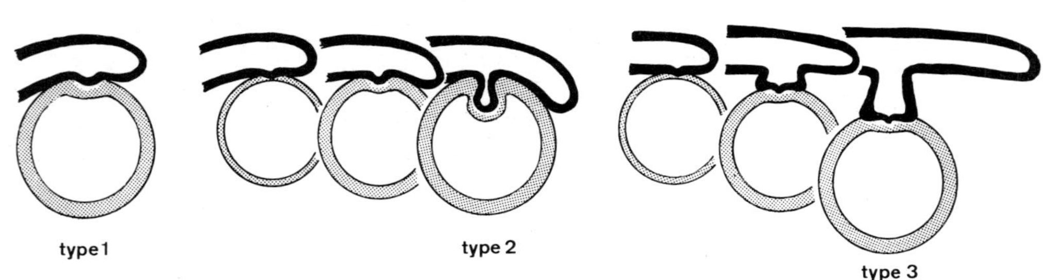

Figure 2. Diagrammatic interpretation of different types of fungus–plant cell interactions in pathogenic and mutualistic symbioses. While intracellular haustoria with a neckband are restricted to rusts and powdery mildews and interparietal haustoria of type 2 and 3 represent a peculiarity of certain groups of lichens, the other types of interactions may be found in different symbiotic systems. Reproduced from Honegger (1986a).

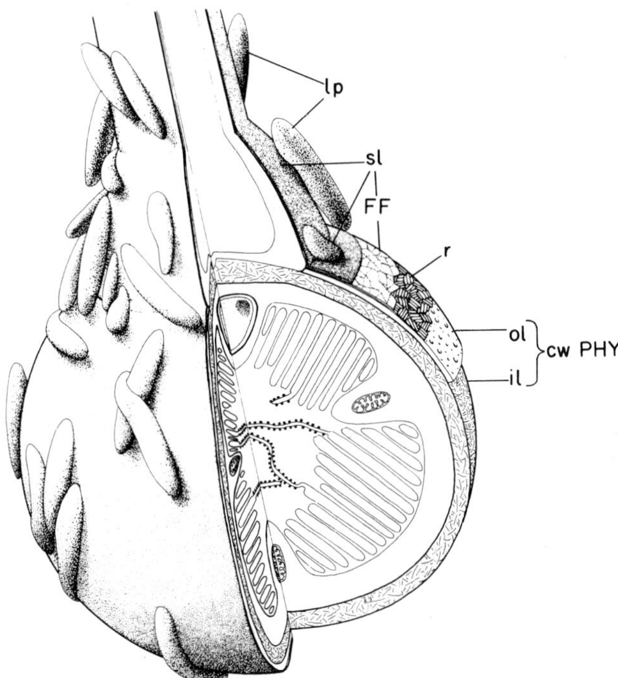

Figure 3. Diagrammatic interpretation of the mycobiont–photobiont interface in Lecanorales with a trebouxioid photobiont and intraparietal type 3 haustoria: lp, crystalline lichen products; sl, common surface layer of mycobiont origin; FF, tesselated surfaces shown in freeze-fracture preparations; r, rodlet layer of unknown origin; al, amorphous outer layer; il, fibrous inner largely cellulosic layer; cw, photobiont cell walls. Reproduced from Honegger (1986b).

variety of levels. At the ecophysiological level, in addition to the shape of the thallus, which optimally exposes the photobiont to appropriate light regimes, the overlying mycobiont tissues may exert physiological roles: acting as a filter for potentially damaging intensities of light by the deposition of pigments and (or) the refractive nature of the cells; accelerating warming by absorption of heat owing to dark pigmentation and vice versa (Kershaw, 1985); reduction of the rate of water loss by means of a thickened, relatively impermeable cortex; regulation and facilitation of gaseous exchange through a mosaic of epicortical pores or other specialized structures ("pseudocyphellae", "cyphellae") (Hale, 1981; Green *et al.*, 1985); and increased retention of minerals, ions, and other compounds, especially in the medullary tissues so that they are available for the metabolism of both bionts when required. While the relative importance of these factors is unexamined in all

ive than in less advanced groups; the increased expense will be proportional to the increased adaptive advantage. (3) Protection from lichen herbivores will be most important in juvenile thalli and reproductive tissues. (4) Total lichen investment in secondary compounds will reflect the cost of replacing resources lost to herbivores; the cost of replacing resources will be a complex function of a lichen's growth rate, reproductive efforts, dispersal ability, and nutrient element content. (5) Investment in lichen secondary compounds will relate to the apparency of lichen thalli to herbivores. Some compounds could be mycotoxic, regulating the development of haustoria by the mycobionts (Ahmadjian, 1987), and further roles proposed are analyzed by Rundel (1978); Wicklow (Chapter 8, this volume) gives further information on some of Lawrey's experiments. Antimicrobial (including antifungal) effects may be especially important (Hawksworth, 1982a), and some compounds form complexes with rock minerals (Purvis *et al.*, 1987) that may have implications for their ability to colonize particular substrata. Further, certain lichen acids may in addition have a role in the control of carbohydrate release from the photobiont through the inhibition of algal urease and a consequent increase in ammonia enhancing cell permeability (Richardson, 1985).

A fascinating observation, the significance of which is not yet understood, is that hyphae from certain bark-dwelling lichens can pass through the bark into the xylem vessels (Orus and Ascaso, 1982; Ascaso, 1985). Metabolites from lichens could therefore be distributed throughout host trees. While evernic acid at least is reported to affect chloroplasts adversely (Ascaso and Rapsch, 1985), the presence of antimicrobial products in tree tissues could conceivably reduce the susceptibility of the tree to attack by pathogenic fungi and bacteria. If such products were established as being translocated to living leaves at concentrations that are microbiologically active, the fitness of both lichen and tree would be increased; coevolution could also have taken place in the establishment of such a beneficial chemical relationship. This work merits more critical study, not least for its possible implications for forest pathology.

The secondary compounds produced by the joint symbioses but not the independent partners are evidently of major adaptive value to the symbiosis. The production of any decarboxylation inhibitor that is effective on the mycobiont is presumably genotypically determined and therefore interpreted as coevolved; it would not have been selected for in the absence of a mycobiont producing the phenolic acid precursors that could then lead to the beneficial secondary metabolites. The mycobiont can also be expected to have undergone genotypic selection towards the abundant production of precursors, as interplay with the photobiont's enzyme inhibitor would lead to increased fitness (e.g., a greater probability of the mycobiont not being destroyed by herbivores).

COEVOLVED STRATEGIES

Sexual strategies

Law and Lewis (1983) found that in mutualistic symbioses with substantial periods of coevolution with one partner, sex was normally reduced in the inhabitants but retained in the exhabitants; this trend is clearly evidenced by the scarcity of sexual reproduction in lichen photobionts (see p. 130). But in lichen symbioses this can be taken a step further, to reduction of sex in the exhabitants as well; this weakens the value of Law and Lewis's (1983) thesis as a general rule. This step is made possible by the production of a range of dual asexual propagules (see p. 133). In many genera, there occur species pairs comprising a regularly sexual species with no asexual propagules and a usually sterile species with abundant asexual propagules (but identical secondary metabolites) (Poelt, 1970). The mainly asexually reproducing ("secondary") species almost invariably have a wider geographical or ecological range than the sexual ("primary") species from which they are postulated to have arisen. This is not surprising, as the production of large numbers of dual propagules facilitates the establishment of new thalli as compatable bionts are dispersed together and the rapid colonization of available ecological niches. In such situations, inability for genetic change does lead to increased fitness, as the genotypes existing are arguably optimally adapted; these combinations are currently generally evolutionarily successful.

"Secondary species" of lichens rarely or never forming sexual stages are often common and widespread. In many cases, probable "primary species" counterparts appear not now to be extant, and their genetic line would therefore have been destined to extinction had no asexual "secondary species" arisen. "Secondary species" are essentially dual clones, with single genotypes for both mycobiont and photobiont. With no possibilities of recombination, "secondary species" must represent evolutionary dead-ends, outstandingly successful at present but as far as we know unable to adapt to future environmental and (or) ecological changes. Phenomena such as parasexuality, aneuploids, or repeat DNA sequences have, however, yet to be searched for amongst lichen-forming taxa.

Niche exploitation

The lichens provide remarkable examples of niche exploitation. In hot and cold deserts and tundra they can become the dominant form of autotrophic life. Ground cover of over 80% by lichens is not uncommon in the boreal zone and they are the most numerous group in the Antarctic flora. In more mesic regions they are able to occupy substrata otherwise inaccessible to most other organisms, for example the surfaces of trunks of living trees, rocks, and buildings, and also air-space in the case of long, pendulous lichens such as species of *Alectoria*, *Bryoria*, *Ramalina*, *Roccella* and *Usnea*.

The production of lichen symbioses is consequently clearly mutualistic in that it increases ecological amplitude by enabling both bionts to colonize areas in which neither could survive alone. Suggestions that mycobionts should be regarded as controlled parasites of the photobionts (e.g., Ahmadjian and Jacobs, 1983) are based on cellular considerations and cannot be sustained in terms of the evolutionary biology and ecology of the biont populations as a whole. The specialized niche exploitation attained would not have been possible without the coevolution of morphological and physiological attributes (see above). As pointed out by Lewis (1987), "... most lichens are clearly mutualistic for the associations are ecologically obligate, i.e. the fitness of both partners approaches zero when they are not together."

Longevity

Crustose lichen symbioses are reportedly amongst the longest-lived recorded, with ages of some individual thalli claimed to be up to 4500 years (Beschel, 1961). While some doubt surrounds such extreme estimates, it is incontravertible that many crustose thalli can persist and continue to produce diaspores over several centuries (e.g., Winchester, 1988). Individual mycobiont hyphae may well survive these extended periods within such colonies, and there are indications of a certain amount of cell turnover in the photobiont (Hill, 1985). Secondary metabolites produced only by the symbiosis may be of special importance in eliminating or reducing threats from potentially parasitic microfungi and grazing invertebrates through such extended periods of time (see p. 140).

This lifestyle contrasts markedly with the situation in most non-lichenized fungi, where the diaspore-producing bodies (sexual or asexual) are relatively ephemeral and generally thrive for some hours to weeks; teleomorphs in a few fungi are perennial (e.g., *Daldinia*, *Ganoderma*, *Patellaria*), but with life-spans at most measured in tens of years.

Longevity in the majority of lichens enables single thalli to continue to produce diaspores of one or both bionts over exceptionally long periods, thereby increasing their chances of finding appropriate partners (where necessary) and new sites in which to develop; that is, it enhances their evolutionary fitness ecologically and by increasing opportunities for experimentation in a new habitat or with new partners.

Coevolution with other organisms

A variety of vertebrates (Richardson and Young, 1977) and a wide range of invertebrates (Gerson and Seaward, 1977) are dependent on lichens for either food or camouflage. It seems probable that some invertebrate lichen-feeders also have a role in the dispersal of lichen fragments or specialized dual propagules, but critical work on such cases is lacking. Nevertheless, it seems probable that examples occur parallel to those described between vertebrates and fungi (Pirozynski and Malloch, this volume, Chapter 10). In the case of reindeer and caribou, however, advantage to the lichen could be just by trampling increasing numbers of thallus fragments available to be blown and splashed onto fresh surfaces.

At least some of the secondary metabolites produced by lichens may be of particular relevance in the coevolution of other organisms with lichens. Antibiotics, antifungal substances and substances damaging to invertebrates have been demonstrated (Lawrey, 1984). These appear to deter potentially harmful organisms and parasites; and commensals of lichens are generally of taxa that occupy only that niche, as in the case of non-parasitic lichenicolous fungi (Hawksworth, 1982a). The possibility that at least some specialized commensalistic fungi may be of unknown benefit to the symbioses has already been alluded to. Further, the evolution of some extant lichens may have involved a parasitic stage on another lichen; the ability of some commensalistic lichenicolous fungi to form galls on their hosts may also be pertinent here.

EVOLUTIONARY SIGNIFICANCE

The evolution of lichen symbioses was interpreted by Bowler and Rundel (1975) as comprising three phases: (1) lichenization; (2) evolution of chemical variation; and (3) the evolution of dual asexual propagules. While their thesis holds good for many of the more familiar lichenized genera, it is not universal. About 20 genera exist that include both lichenized species and ones with other nutritional methods (Hawksworth and Hill, 1984), some of which are drawn from groups of fungi that are unlikely to have been anciently lichenized (e.g., Agaricales, Dothideales, Helotiales). I consider that experimentation and selection are in progress in such genera, while in others more related to the major lichenized groups, for example Arthoniales or Lecanorales, a switch from lichen-forming to lichen parasites, commensals, or saprobes may be occurring. The result is a complex network of evolving and devolving lichen associations representing a wide range of fungus–alga/cyanobacterial relationships (Hawksworth, 1978).

As pointed out by Hawksworth (1988a), studies relating the distribution of lichen families and genera to continental movements are increasingly pointing to a considerable antiquity for certain groups of lichens. In some cases, genera, species, and even chemical races might have evolved by the Permo-Triassic. Further support for the antiquity of some lichenized orders can also be obtained from a study of the obligately lichenicolous fungi (Hawksworth, 1982a; Eriksson and Santesson, 1986), and from comparison with the ascal structure of the host lichens (Eriksson, 1981; Hawksworth, 1982b). Ascomycetes may have already been quite diverse by the Silurian (Sherwood-Pike and Gray, 1985), and this view is consistent with that possibility. Cyanobacteria themselves appear to be much more ancient than green algae, extending perhaps to the Proterozoic, 2.5 thousand million years ago (Schopf and Walter, 1982).

While symbioses involving cyanobacteria have sporadically evolved *de novo* in many groups of plants and some animals (Smith and Douglas, 1987), the avaliable indications are that the coevolution of fungi with cyanobacteria preceded that with green algae. This is suggested by the probably ancestral ascus structures of many such lichens, especially the Peltigerales (Hawksworth, 1982b). Lichenization has evidently been a key feature in the early evolution of the ascomycetes. Orders and families now opting for other methods of nutrition, for example as saprobes on wood, mammal dung, parasites of flowering plants, or mycorrhizas with phanerogams, would not have had these nutritional possibilities available when the group was diversifying prior to the Permian. The production of stromata has also been considered an ancestral ascomycete feature (Corner, 1964); stromata are optimally developed in foliose and fruticose lichenized stromata (i.e., lichen thalli). Ancestral traits in ascomycetes have therefore to be searched for in widely distributed lichenized groups occurring particularly on rocks and on the ground. The lichen-forming fungi cannot therefore be ignored in considerations of the evolution of the Ascomycotina (and their anamorphs) as a whole.

ACKNOWLEDGEMENTS

I am indebted to Dr D. M. John and Dr B. A. Whitton for advice on the placement of green and blue-green photobionts; to Dr K. A. Pirozynski for stimulating discussions and critical comments on my draft manuscript; to Dr R. Honegger for permission to include Figures 2 and 3; and to Mrs M. S. Rainbow for word processing.

REFERENCES

Ahmadjian, V. (1962). Investigations on lichen synthesis. *Am. J. Bot.* **49,** 277–283.
Ahmadjian, V. (1967). "The Lichen Symbiosis". Blaisdell, Waltham, Mass.

Ahmadjian, V. (1970). The lichen symbiosis, its origin and evolution. *Evol. Biol.* **4,** 163–184.

Ahmadjian, V. (1982). Algal/fungal symbioses. In "Progress in Physiological Research" (F. E. Round and R. L. Chapman, eds), vol. 1, pp. 179–233. Elsevier Biochemical Press, Amsterdam.

Ahmadjian, V. (1987). Coevolution in lichens. *Ann. N.Y. Acad. Sci.* **503,** 307–315.

Ahmadjian, V. and Hekkilä, H. (1970). The culture and synthesis of *Endocarpon pusillum* and *Staurothele clopima. Lichenologist* **4,** 259–267.

Ahmadjian, V. and Jacobs, J. (1983). Algal–fungal relationships in lichens: recognition, synthesis, and development. In "Algal Symbiosis" (L. J. Geoff, ed.), pp. 147–172. Cambridge University Press, Cambridge.

Ahmadjian, V. and Jacobs, J. (1985). Artificial reestablishment of lichens. IV. Comparison between natural and synthetic thalli of *Usnea strigosa. Lichenologist* **17,** 149–165.

Ahmadjian, V., Russell, L. A., and Hildreth, K. C. (1980). Artificial reestablishment of lichens. I. Morphological interactions between the phycobionts of different lichens and mycobionts *Cladonia cristella* and *Lecanora chrysoleuca. Mycologia* **72,** 73–89.

Ahti, T. (1982). The morphological interpretation of cladoniiform thalli in lichens. *Lichenologist* **14,** 105–113.

Ascaso, C. (1985). Structural aspects of lichens invading their substrata. In "Surface Physiology of Lichens" (C. Vicente, D. H. Brown, and M. E. Legaz, eds), pp. 87–113. Universidad Complutense de Madrid, Madrid.

Ascaso, C. and Rapsch, S. (1985). Effect of evernic acid on structure of spinach chloroplasts. *Ann. Bot.* **56,** 467–473.

Beschel, R. E. (1961). Dating rock surfaces by lichen growth and its application to glaciology and physiography (lichenometry). In "Geology of the Arctic" (G. D. Raasch, ed.), vol. 2, pp. 1044–1062. University of Toronto Press, Toronto.

Bourrelly, P. (1973). "Les Algues d'Eau Douce. Initiation à la Systematique" Vol. 1. "Les Algues Vertes". Revised edn. Boubee, Paris.

Bowler, P. A. and Rundel, P. W. (1975). Reproductive strategies in lichens. *Bot. J. Linn. Soc.* **70,** 325–340.

Brunner, U. and Honegger, R. (1985). Chemical and ultrastructural studies on the distribution of sporopollenin-like biopolymers in six genera of lichen phycobionts. *Can. J. Bot.* **63,** 2221–2230.

Bubrick, P. and Galun, M. (1986). Spore to spore resynthesis of *Xanthoria parietina. Lichenologist* **18,** 47–49.

Bubrick, P., Galun, M., and Frensdorf, A. (1981). Proteins from the lichen *Xanthoria parietina* which bind to phycobiont cell walls. *Protoplasma* **105,** 207–211.

Cain, R. F. (1972). Evolution of the fungi. *Mycologia* **64,** 1–14.

Corner, E. J. H. (1964). "The Life of Plants". Weidenfeld and Nicolson, London.

Culberson, C. F. and Ahmadjian, V. (1980). Artificial reestablishment of lichens. II. Secondary products of resynthesized *Cladonia cristatella* and *Lecanora chrysoleuca. Mycologia* **72,** 90–109.

Culberson, C. F., Culberson, W. L., and Johnson, A. (1977). "Second Supplement to 'Chemical and Botanical Guide to Lichen Products'". American Bryological and Lichenological Society, St Louis.

Degelius, G. (1954). The lichen genus *Collema* in Europe. *Symb. bot. upsal.* **13**(2), 1–499.

Dick, M. W. and Hawksworth, D. L. (1985). A synopsis of the biology of the Ascomycotina. *Bot. J. Linn. Soc.* **91,** 175–179.

Elix, J. A., Whitton, A. A., and Sargent, M. V. (1984). Recent progress in the chemistry of lichen substances. *Prog. Chem. Org. nat. Prod.* **45,** 103–234.

Eriksson, O. (1981). The families of bitunicate ascomycetes. *Opera bot.* **60,** 1–220.

Eriksson, O. and Santesson, R. (1986). *Lasiosphaeriopsis stereocaulicola. Mycotaxon* **25,** 569–580.

Fahselt, D. (1985). Multiple enzyme forms in lichens. In "Lichen Physiology and Cell Biology" (D. H. Brown, ed.), pp. 129–143. Plenum Press, New York and London.

Friedl, T. (1987). Thallus development and phycobionts in the parasitic lichen *Diploschistes muscorum. Lichenologist* **19,** 183–191.

Gerson, U. and Seaward, M. R. D. (1977). Lichen-invertebrate associations. In "Lichen Ecology" (M. R. D. Seaward, ed.), pp. 69–119. Academic Press, London and New York.

Green, T. G. A., Snelgar, W. P., and Wilkins, A. L. (1985). Photosynthesis, water relations and thallus structure of Stictaceae lichens. In "Lichen Physiology and Cell Biology" (D. H. Brown, ed.), pp. 57–75. Plenum Press, New York and London.

Greenhalgh, G. N. and Whitfield, A. (1987). Thallus tip structure and matrix development in *Bryoria fuscescens. Lichenologist* **19,** 295–305.

Hafellner, J. and Poelt, J. (1980). Der Flechtenparasite *Buellia pulverulenta*-eine bleinbend interneparasitische Flechten. *Phyton, Horn* **20,** 129–133.

Hale, M. E. (1981). Pseudocyphellae and pored epicortex in the Parmeliaceae, their delimitation and evolutionary significance. *Lichenologist* **13,** 1–10.

Hawksworth, D. L. (1978). The taxonomy of lichen-forming fungi, reflections on some fundamental problems. In "Essays in Plant Taxonomy" (H. E. Street, ed.), pp. 211–243. Academic Press, London.

Hawksworth, D. L. (1982a). Secondary fungi in lichen symbioses: parasites, saprophytes and parasymbionts. *J. Hattori Bot. Lab.* **52,** 357–366.

Hawksworth, D. L. (1982b). Co-evolution and the detection of ancestry in lichens. *J. Hattori Bot. Lab.* **52,** 323–329.

Hawksworth, D. L. (1985). Problems and prospects in the systematics of the Ascomycotina. *Proc. Indian Acad. Sci.* (*Pl. Sci.*) **94,** 319–339.

Hawksworth, D. L. (1988a). The variety of fungal-algal symbioses, their evolutionary significance, and the nature of lichens. *Bot. J. Linn. Soc.* **96,** 3–20.

Hawksworth, D. L. (1988b). The fungal partner. In "Handbook of Lichenology" (M. Galun, ed.), vol. 1. CRC Press, Boca Raton, Fla., in press.

Hawksworth, D. L. and Hill, D. J. (1984). "The Lichen-Forming Fungi". Blackie, Glasgow and London.

Hawksworth, D. L. and Poelt, J. (1986). Five additional genera of conidial lichen-forming fungi from Europe. *Pl. Syst. Evol.* **154,** 195–211.

Hawksworth, D. L., Sutton, B. C., and Ainsworth, G. C. (1983). "Ainsworth & Bisby's Dictionary of the Fungi", 7th edn. Commonwealth Mycological Institute, Kew.

Hildreth, K. C. and Ahmadjian, V. (1981). A study of *Trebouxia* and *Pseudotrebouxia* isolates from different lichens. *Lichenologist* **13,** 65–86.

Hill, D. J. (1985). Changes in photobiont dimensions and numbers during co-development of lichen symbionts. In "Lichen Physiology and Cell Biology" (D.H. Brown, ed.), pp. 303–317. Plenum Press, New York and London.

Honegger, R. (1984). Cytological aspects of the mycobiont-phycobiont relationship in lichens. *Lichenologist* **16,** 111–127.

Honegger, R. (1986a). Ultrastructural studies in lichens I. Haustorial types and their frequencies in a range of lichens with trebouxioid photobionts. *New Phytol.* **103,** 785–795.

Honegger, R. (1986b). Ultrastructural studies in lichens II. Mycobiont and photobiont surface layers and adhering crystalline lichen products in four Parmeliaceae. *New Phytol.* **103,** 797–808.

Honegger, R. (1987). Questions about pattern formation in the algal layer of lichens with stratified (heteromerous) thalli. *Biblthca lich.* **25,** 59–71.

Honegger, R. and Brunner, U. (1981). Sporopollenin in the cell walls of *Coccomyxa* and *Myrmecia* phycobionts of various lichens: an ultrastructural and chemical investigation. *Can. J. Bot.* **59,** 2713–2734.

Jahns, H. M. (1974) ["1973"]. Anatomy, morphology, and development. In "The Lichens" (V. Ahmadjian and M. E. Hale, eds), pp. 3–58. Academic Press, New York.

Jahns, H. M. (1988). The establishment, individuality and growth of lichen thalli. *Bot. J. Linn. Soc.* **96,** 21–29.

James, P. W. and Henssen, A. (1976). The morphological and taxonomic significance of cephalodia. In "Lichenology: Progress and Problems" (D. H. Brown, D. L. Hawksworth, and R. H. Bailey, eds), pp. 27–77. Academic Press, London.

Kershaw, K. A. (1985). "Physiological Ecology of Lichens". Cambridge University Press, Cambridge.

Kohlmeyer, J. and Kohlmeyer, E. (1972). Is *Ascophyllum nodosum* lichenized? *Bot. mar.* **15,** 109–112.

Lallement, R. and Savoye, D. (1985). Lectins and morphogenesis: facts and outlook. In "Lichen Physiology and Cell Biology" (D. H. Brown, ed.), pp. 335–350. Plenum Press, New York and London.

Lallement, R., Boissière, M. -C., Le Clerc, J. -C., Velly, P., and Wagner, J. (1987). La symbiose lichénique: approaches nouvelles. *Bull. Soc. bot. Fr.* **137,** *Act. bot.* 1986-2, 41–79.

Law, R. (1985). Evolution in a mutualistic environment. In "The Biology of Mutualism" (D. H. Boucher, ed.), pp. 145–170. Croom Helm, London and Sydney.

Law, R. and Lewis, D. H. (1983). Biotic environments and the maintenance of sex—some evidence from mutualistic symbioses. *Biol. J. Linn. Soc.* **20,** 249–276.

Lawrey, J. D. (1984). "Biology of Lichenized Fungi". Praeger Scientific, New York.

Lewis, D. H. (1973). The relevance of symbiosis to taxonomy and ecology, with particular reference to mutualistic symbiosis and the exploitation of marginal habitats. In "Taxonomy and Ecology" (V. H. Heywood, ed.), pp. 151–172. Academic Press, London and New York.

Lewis, D. H. (1987). Evolutionary aspects of mutualistic associations between fungi and photosynthetic organisms. In "Evolutionary Biology of the Fungi" (A. D. M. Rayner, C. M. Brasier, and D. Moore, eds), pp. 161–178. Cambridge University Press, Cambridge.

Marton, K. and Galun, M. (1976). *In vitro* disassociation and reassociation of the symbionts of the lichen *Heppia echinulata*. *Protoplasma* **87,** 135–143.

Mollenhauer, D. (1970). Botanische Notizen Nr. 1: Beobachtungen au der Blaualge *Geosiphon pyriforme*. *Natur und Museum* **100,** 213–223.

Molnar, K. E., Stewart, K. D., and Mattox, K. R. (1975). Cell division in the filamentous *Pleurastrum* and its comparison with the unicellular *Platymonas* (Chlorophyceae). *J. Phycol.* **11,** 287–296.

Moreau, F. (1956). Sur la théorie biomorphogénique de lichens. *Revue bryol. lichén.* **25,** 183–186.

Moser-Rohrhofer, M. (1975). "Physiologische und vergleichende Anatomie der Flechtenpilze". Vol. 1. Akademische Druck-um Verlagsanstalt, Graz.

Orus, I. and Ascaso, C. (1982). Localization de hifas liquencias en lostejidos conductores de *Quercus rotundiifolia* Lam. *Collnea Bot.* **13,** 325–338.

Ott, S. (1987a). Reproductive strategies in lichens. *Biblthca lich.* **25,** 81–93.

Ott, S. (1987b). Sexual reproduction and developmental adaptations in *Xanthoria parietina*. *Nordic J. Bot.* **7,** 219–228.

Parlevliet, J. E. (1986). Coevolution of host resistance and pathogen virulence, possible implications for taxonomy. In "Coevolution and Systematics" (A. R. Stone and D. L. Hawksworth, eds), pp. 19–34. Clarendon Press, Oxford.

Piattelli, M. and de Nicola, M. G. (1968). Anthraquinone pigments from *Xanthoria parietina* (L.). *Phytochemistry* **7,** 1183–1187.

Poelt, J. (1970). Das Konzept der Artenpaare bei den Fechten. *Vortr. bot. Ges.*, n.f. **4,** 187–198.

Poelt, J. (1985). *Caloplaca epithallina*. Porträt einer parasitischen Flechte. *Bot. Jb.* **167,** 457–468.

Purvis, O. W., Elix, J. A., Broomhead, J. A., and Jones, G. C. (1987). The occurrence of copper-norstictic acid in lichens from cupriferous substrata. *Lichenologist* **19,** 193–203.

Richardson, D. H. S. (1985). The surface physiology of lichens with particular reference to carbohydrate transfer between the symbionts. In "Surface Physiology of Lichens" (C. Vicente, D. H. Brown, and M. E. Legaz, eds), pp. 25–55. Universidad Complutense de Madrid, Madrid.

Richardson, D. H. S. and Young, C. M. (1977). Lichens and vertebrates. In "Lichen Ecology" (M. R. D. Seaward, ed.), pp. 121–144. Academic Press, London and New York.

Rundel, P. W. (1978). The ecological role of secondary lichen substances. *Biochem. Syst. Ecol.* **6,** 157–170.

Santesson, R. (1952). Foliicolous lichens I. A revision of the taxonomy of the obligately foliicolous lichenized fungi. *Symb. bot. upsal.* **12**(1), 1–590.

Schopf, J. W. and Walter, M. R. (1982). Origins and early evolution of cyanobacteria, the geological evidence. In "The Biology of Cyanobacteria" (N. G. Carr and B. A. Whitton, eds), pp. 543–564. University of California Press, Berkeley and Los Angeles.

Sherwood-Pike, M. A. and Gray, J. (1985). Silurian fungal remains: probable records of the class Ascomycetes. *Lethaia* **18,** 1–20.

Slocum, R. D. (1980). Light and electron microscopic investigations in the Dictyonemataceae (basidiolichens). II. *Dictyonema irpicinum*. *Can. J. Bot.* **58,** 1005–1015.

Slocum, R. D., Ahmadjian, V., and Hildreth, K. C. (1980). Zoosporogenesis in *Trebouxia gelatinosa*: Ultrastructure potential for zoospore release and implications for the lichen association. *Lichenologist* **12,** 173–187.

Smith, D. C. and Douglas, A. E. (1987). "The Biology of Symbiosis". Edward Arnold, London.

Steiner, M. and Poelt, J. (1987). Die parasitische Flechten auf *Caloplaca polycarpoides*. *Pl. Syst. Evol.* **155,** 133–141.

Stewart, W. D. P., Rowell, P., and Rai, A. N. (1983). Cyanobacteria–eukaryotic plant symbioses. *Ann. Microbiol.* **134B,** 205–228.

Stocker-Wörgötter, E. and Türk, R. (1987). Die Resynthese der Flechte *Verrucaria macrostoma* unter Laborbedingungen. *Nova Hedwigia* **44,** 55–68.

Thompson, J. N. (1986). Patterns in coevolution. In "Coevolution and Systematics" (A. R. Stone and D. L. Hawksworth, eds), pp. 119–143. Clarendon Press, Oxford.

Thompson, J. N. (1987). Symbiont-induced speciation. *Biol. J. Linn. Soc.* **32,** 385–393.

Tibell, L. (1982). Caliciales of Costa Rica. *Lichenologist* **14,** 219–254.

Tschermak-Woess, E. (1978). *Myrmecia reticulata* as a phycobiont and free-living—Free-living *Trebouxia*—The problem of *Stenocybe septata*. *Lichenologist* **10,** 69–79.

Uyenco, F. R. (1965). Studies on some lichenized *Trentepohlia* associated in lichen thalli with *Coenogonium*. *Trans. Am. microsc. Soc.* **84,** 1–14.

Vobis, G. and Hawksworth, D. L. (1981). Conidial lichen-forming fungi. In "The Biology of Conidial Fungi" (G. T. Cole and B. Kendrick, eds), pp. 245–273. Academic Press, New York.

Ward, S. M. (1884). On the structure, development, and life-history of a tropical epiphyllous lichen (*Strigula complanata*, Fée, fide Rev. J. M. Crombie). *Trans. Linn. Soc. Lond.*, ser. 2, Bot. **2,** 87–119.

Winchester, V. (1988). An assessment of lichenometry as a method for dating recent stone movements in two stone circles in Cumbria and Oxfordshire. *Bot. J. Linn. Soc.* **96,** 57–68.

7 | Coevolution of Entomogenous Fungi and their Insect Hosts

H. C. EVANS

CAB International Institute of Biological Control, Ascot, Berkshire, UK

Abstract

Four groups of entomogenous fungi are delimited: entomopathogens, mutualistic symbionts, ectoparasites, and endoparasites. Much of the available evidence for coevolution is overwhelmingly circumstantial and relates to entomopathogens, to which the major part of this chapter is devoted. After defining terms, aspects of insect defence mechanisms and how these are overcome by the various fungi involved are considered. The biology and phylogeny of these fungi are discussed with reference to their hosts, and it is concluded that fungal parasitism arose *de novo* in the major subdivisions of the fungi. The possible origins of the entomogenous habit are speculatively examined.

INTRODUCTION

An introduction should acquaint the reader with past and present knowledge of the subject under discussion. Such an introduction is superfluous here, since the theme of coevolution has scarcely been touched upon in entomogenous fungi, as Humber (1984) reflected for the Entomophthorales: ". . . too little is known of the host spectrum or the factors controlling the host specificities . . . to make anything more than general comments about the possible coevolution of these fungi with their insect hosts." Consequently, it is a largely unexplored area where one should tread warily. Nevertheless, speculative hypotheses must be attempted and made available for testing if progress is to be more than pedestrian. Speculation has been urged by the editors of this book; I have accepted their challenge. The facts are scarce and the evidence, when available, is circumstantial and thus controversial. I therefore make my apologies in advance.

I begin with controversy in trying to define the terms of the subject entrusted to me. As Luttrell (1974) perceptively remarked in discussing the terminology of parasitism: "In the best taxonomic tradition I seek to bring

COEVOLUTION OF FUNGI
WITH PLANTS AND ANIMALS
ISBN 0-12-557365-0

order by synonymizing names proposed by others while proposing new ones of my own." Firstly, coevolution—what exactly is it within the present context? Parlevliet (1986), in a treatise devoted mainly to fungal pathogens of plants, discusses at length the concept of coevolution, interpreting it as "... mutual evolution: the host evolving defence mechanisms, the parasite evolving mechanisms to circumvent or neutralize these defences" (see, however, Pirozynski and Hawksworth, this volume, Chapter 1). At the cellular level or microlevel, therefore, coevolution is self-explanatory. Thus, here I shall discuss aspects of pathogenesis: the mechanisms that insects have evolved to prevent invasion by fungi and the corresponding increase in sophistication of fungal pathosystems to break down these defence barriers. As with plant pathogens, this has resulted in extremes of host specialization: the hallmark of coevolution. Coevolution is also regarded by many as the adaptation of the parasite to the life cycle of the host in order to survive and, for insects, to behavioural as well as seasonal patterns. This broad interpretation of coevolution at the macrolevel is adopted here and I attempt to introduce examples within the entomogenous fungi.

The term entomogenous is applied here only to those fungi which have developed a parasitic association with insects. The fungi involved in intimate but non-parasitic relationships with insects are dealt with in other chapters. If the climax of coevolution is the establishment of a mutualistic symbiosis (Parlevliet, 1986), then examples are rare amongst the entomogenous fungi. However, the spectre of terminology looms large here. Hawksworth *et al.* (1983) define symbiosis as the living together of unlike organisms, but add the rider that current usage is much more limited, in particular referring only to reciprocal parasitism. This is the popular image of symbiosis as embodied in most dictionaries: "a comradeship; association of two different organisms [which] contribute to each others' support" (Oxford English Dictionary, 1971). However, de Bary (1879) used symbiosis to cover any living association, no matter how transient, and this all-embracing concept is now generally accepted (Cooke, 1977; Lewis, 1985; Smith and Douglas, 1987). Cooke (1977) includes entomogenous fungi within the symbiotic fungi and delimits three groups: neutral obligate symbionts, antagonistic endosymbionts, and mutualistic symbionts. This terminology will not be considered further in order to "... be free of ambiguity ... and to avoid ... the current proliferation of terms" (Cooke, 1977). Clearly, this is unsafe territory and I will proceed no further into this semantic quagmire (Lewis, 1985). Suffice it to say that in the historical usage, pre-dating de Bary, of the term symbiosis an interdependence was implicit. This ultimate expression of coevolution is found amongst entomogenous taxa only in certain Septobasidiales (entomogenous mutualistic symbionts). In the majority of entomogenous fungi, there is a one-way transfer of nutrients or benefits and it is possible to recognize three groups: entomogenous ectoparasites, entomogenous endoparasites, and entomopathogens. The first two are parasites that live externally or internally on insects but that do not appear to affect their hosts deleteriously, whereas the last invade and disrupt the tissues of the host, leading to its death (Samson *et al.*, 1988). I shall discuss these fungi in order of their economic relevance to man, which is directly proportional to the amount of literature on them.

PRELIMINARY CONSIDERATIONS

For most of the fungi included here, the cuticle is either the gateway to or the site of their parasitism. The exoskeleton is a formidable barrier composed of a complex of layers. The increasing sophistication of this structure enabled the insects to colonize the most diverse of terrestrial habitats and today they represent over 70% of all known animals. In excess of 800 000 species have been recognized. The first insects probably adapted to terrestrial niches during the early Devonian (*c.* 350 million years ago), and wingless insects have been found in late Devonian sediments. By way of comparison, the first land plants probably evolved during the Silurian, 380 million years ago (Savile, 1955). The habitat was moist and tropical, the scene of frenetic evolution in the plant kingdom, and the precursor of the humid tropical forest. The oldest records of an insect–fungal association have been discovered in Miocene–Oligocene amber, aged approximately 25 million years (Poinar and Thomas, 1982, 1984). They have been interpreted as pertaining to present-day entomopathogenic genera (*Beauveria*, *Entomophthora*). This was also the period when modern insect genera were being delimited. It is likely, however, that the forerunners of extant entomogenous fungi were coevolving with the insects from the earliest (Paleozoic) times. Significantly, perhaps, the polysaccharide chitin is common to both arthropods and fungi, occurring in most fungal cell walls and insect cuticles. Chitin, therefore, would have been a frequent component of potentially exploitable substrates amongst the debris of tropical soils; indeed, fossil chitinous organisms (Chitinozoa) have been known since the Precambrian period. Utilization of it by heterotrophs was inevitable. This would have been the first step in the development of the entomogenous mode of life.

THE INSECT DEFENCES INVOLVED (HOST VERSUS PARASITE)

The battle for the cuticle

The chitin–protein cuticle is considered to have been a central development in the evolution of insects (Wigglesworth, 1957). This external barrier is mirrored to a certain extent internally by the chitinous intestinal epithelium that prevents the ingress of microorganisms into the body cavity. However, as nearly all entomogenous fungi seek entry via the exoskeleton, this discussion is restricted to the cuticle. Why more fungi have not adapted to the seemingly easier endoparasitic habit is interesting, as this was the original concept of how entomopathogens initiated infection (Cooke, 1892).

The cuticle is composed of three non-cellular layers above the living epidermis. The outer layer ("epicuticle") is composed of lipoproteins and wax: a hard, impervious, inhospitable structure that in addition contains antifungal substances, particularly epicuticular lipids, that differ qualitatively and quantitatively between insect species and may be toxic to germinating fungal spores (Smith and Grula, 1982). This chemical and physical barrier would have deterred the early chitinoclastic saprophytic fungi. As the ancestors of entomogenous fungi overcame this toxicity, novel fatty acids and new combinations of existing ones would have evolved. The host versus parasite conflict had commenced. The first steps towards increasing specialization of the parasitic habit were at the epicuticular level. Recognition of the host by the parasite would also have been an important evolutionary advance, and the spore germination of some entomopathogens is stimulated by epicuticular extracts (Charnley, 1984). Unfortunately, investigations so far have been limited to common species with broad host ranges; more complex early recognition mechanisms can be expected in more host-specific parasites. For example, some aquatic entomopathogens produce motile spores that are chemotactically attracted to their hosts, a relatively sophisticated adaptation analagous to those of several groups of plant pathogens. In addition to identifying the potential host epicuticle and overcoming any fungal toxins, the parasite first has to land on and adhere to the host. This involves the development of diaspores designed to hit a mobile target or to ensure that insects come into contact with them. Methods of fungal dispersal are considered below in relation to the fungi involved.

Once host contact is achieved, spore morphology is of paramount importance in ensuring successful attachment and subsequent infection. Only recently has this aspect been investigated in any depth and solely for the entomopathogens. As Boucias and Latge (1986) pointed out, spore attachment involves a complex of non-specific (electrostatic) and specific mechanisms. Many of the Entomophthorales produce spores in an amorphous mucus, and slimy conidia are commonly formed by anamorphs of the Clavicipitales, providing an adaptation for adhering to insect cuticles. Eilenberg *et al.* (1986) investigated the ultrastructure of the spores of several Entomophthorales and demonstrated that the wall is two-layered and, in addition, that the outer is laminated and ruptures on impact to release mucilaginous material. Dry spores, such as the ascospores of Clavicipitales and most hyphomycete conidia, appear to be unsuited for adhesion to waxy epicuticle. However, the dry spores of some entomopathogenic Hyphomycetes are apparently covered by fascicles of rodlets, their pattern and arrangement varying with the fungus. The rodlets may function together with glycoproteins (lectins) in binding the spores to the cuticle. Such is the tenacity of the binding that attempts to remove dry spores of various entomopathogens by treatment with boiling solvents have proved unsuccessful (Boucias and Latge, 1986; Samson *et al.*, 1988). The increasing sophistication of spore morphology allows for no reciprocal adaptation by the host, although insect grooming may be a direct response to impaction of fungal spores on the cuticle.

Once on the cuticle, the fungus has to breach the outer defences. The infection process has been studied for relatively few entomogenous fungi, but appressorial formation appears to be a common phenomenon, at least in the Ascomycotina and their anamorphs. Release of extracellular lipases from the appressorium may improve adhesion and also provide nutrients, leading to further enzyme induction, especially of proteases and later of chitinases. All these enzymes have been identified in cultures or extracts of entomogenous fungi (Roberts and Humber, 1984). After penetration of the epicuticle, the fungal mycelium or infection peg has to penetrate a complex laminated procuticle, consisting of exocuticle (chitin, protein, phenols) and endocuticle (chitin, protein). Further penetration is apparently slow and consists of a stepwise physical rupture and enzymatic dissolution of the procuticular laminae. Each group of entomogenous fungi has probably evolved different combinations of enzymes and hyphal structures to achieve this end. On reaching the epidermis, therefore, the fungal parasite has overcome an impressive array of outer defences; few other organisms have developed this capability. It has had to evolve strategies to adhere to and germinate on the insect cuticle, detoxify antifungal substances in the epicuticle, degrade host-specific lipids, proteins, and chitin, and rupture the complex laminated glycoproteins in the procuticle. Unquestionably, this indicates a long period of association and coevolution of host and parasite; only the most highly adapted fungi have succeeded. The ectoparasites stop here; apparently, they can obtain sufficient nutrients in this position, via sophisticated holdfast-haustorial structures, to satisfy their sporulation needs. The entomopathogens and mutualistic symbionts require methods of overcoming inner defences.

The battle for the body

Gunnarsson (1987) compared the penetration and slow colonization of the cuticular layers to a build-up of forces in order to establish a stronghold from which to launch an attack against the cellular and humoral defences of the host. Insects have evolved an immune response through haemocytes that involves phagocytosis, nodule formation, and encapsulation in order to identify and destroy the invading organisms. Haemocyte activity occurs in the haemocoele, but there is evidence of various humoral immunity factors located in the epidermis and fat bodies. Gunnarsson (1987) has shown that haemocytes gather around the site of infection and even penetrate the epidermis. Melanization, with the release of prophenoloxidases, also occurs in the procuticle and spreads into the epidermal cells. Although these defences are rapidly and massively deployed, entomopathogenic fungi manage to invade the haemocoele, apparently by disrupting haemocyte activity. Haemocytes aggregating around the hyphae show abnormal responses and altered morphology, possibly due to the release of toxins by the fungus, and prove to be incapable of ingesting, or otherwise preventing invasion by, hyphae. Colonization of the haemocoele is thus unchecked. In less pathogenic or more highly adapted parasites or mutualists, there is evidence of minimal host damage and an indication that the fungus misleads the inner defences into not recognizing it as an intruder: a failure to distinguish between self and non-self, which is the high point of host–parasite coevolution. The fungi that reach and colonize the insect body via the cuticle are highly evolved parasites, having adapted to and neutralized the entire range of physical, chemical, and cellular defences that insects have deployed to resist such attacks.

THE FUNGI INVOLVED

Entomopathogens

Entomopathogens are found amongst the Mastigomycotina, Zygomycotina, and Ascomycotina. The Septobasidiales (Basidiomycotina) can penetrate and grow within insects, but the mode of colonization and nutrition is distinct from that of the fungi described below. Invasion of the insect haemocoele results in a disease syndrome that sooner or later ends with death of the host. These fungi, therefore, are obligately pathogenic and, because of the ubiquity of certain orders on agriculturally important insect pests, they have been the subject of much interest. Consequently, I shall confine this discussion mainly to these well-researched orders in the Zygomycotina (Entomophthorales), Ascomycotina (Clavicipitales, Hypocreales) and hyphomycetous anamorphs.

Zygomycotina

Nearly all the entomopathogenic members belong to the Entomophthorales. Humber (1984) examined evolution within this order with particular reference to the mode of life of ancestral Entomophthorales. His conclusions concur with the view that parasitism arose from saprotrophy, and not vice versa. Circumstantial evidence supports the suggestion that all genera of Entomophthorales radiated from a terricolous, saprophytic habit. *Conidiobolus* may represent the most ancient of the extant genera in the order, and precursors of the Entomophthorales were probably similar to species of this genus that grew saprophytically in the wet, warm, Devonian soils (350 million years ago). These extinct forms might have evolved enzyme systems to exploit insect cadavers in competition with other heterotrophs. Parasitism could then have inevitably followed as a refuge from competing saprophytes. Step-by-step adaptation may have occurred, firstly to colonize predisposed (damaged or otherwise debilitated) insects, and then to invade and kill healthy individuals, possibly apterous, relatively slow-moving, and vulnerable targets. Some extant species of *Conidiobolus*, such as *C. coronatus*, are designated as secondary or opportunistic pathogens by Papierok (1986). *C. coronatus* can overgrow and replace more specialized Entomophthorales on insects in the tropics. There is no evidence of host specificity in these facultatively pathogenic *Conidiobolus* species, and they can probably attack any stressed insect. Compare this with obligately pathogenic species, such as *C. obscurus*, which is restricted to Aphididae, and *C. conglomeratus*, parasitizing only culicid flies. Moreover, the so-called saprophytic species are characterized by rapid host kill; colonization of the host occurs only after death. Toxins have been identified in culture filtrates of these taxa, but appear to be absent in more specialized or host-specific species (Humber, 1984; Papierok, 1986). As Papierok concluded, in *Conidiobolus* there is a range of adaptive mechanisms that demonstrate that insect and fungus have coevolved over a considerable period of time. At first sight, it could be concluded that the development of a toxin that rapidly immobilizes and kills the host is evolutionarily advanced. However, if toxin production is not accompanied by additional metabolites that ensure that the cadaver remains free from competing, opportunistic microorganisms, then the pathogen is at a severe disadvantage. Bacteria often displace these toxin-producing Entomophthorales and septicaemia sets in before hyphal colonization is complete. If, as it appears, the Entomophthorales never evolved antibiotic systems, then the evolutionary option was to keep the host alive as long as possible: essentially to colonize the living insect and claim sufficient substrates to maximize sporulation whilst at the same time reducing the danger of being over-run by microbial competitors. Increased parasitism, reduced pathogenicity (by restricted growth in and colonization of the host) and increased host specialization are characteristic of the most highly evolved species and genera of the Entomophthorales (Humber, 1982, 1984). The saprophytic habit has been lost in such fungi and the entire life cycle conforms to that expected of a biotroph (Luttrell, 1974; Cooke, 1977). Species of *Strongwellsea*, *Massospora*, and some of *Entomophthora* grow and sporulate only in living insects (Humber, 1976; Soper *et al.*, 1976; Ben-Ze'ev *et al.*, 1985). These fungi are correspondingly highly host-specific. All species of *Massospora* for example, are restricted to cicada hosts, whilst *S. castrans* infects only dipterans in the *Hylemya* complex (Anthomyiidae). The latter fungus, once inside the host, is mainly restricted to the haemocoele of the abdomen, although hyphae may also penetrate the nervous system, eventually reaching the brain tissues. There appears to be no tissue damage, however, and no marked change in host behaviour; the insect dies not from histolysis or lethal toxins but owing to a fungal-induced starvation (Humber, 1982). As the author concludes, "... *Strongwellsea* is on an evolutionary course that will progressively restrict vegetative growth and establish a parasitic rather than a pathogenic relationship with its hosts." He further notes that both this genus and *Massospora* have diverged from the mainstream of the Entomophthorales into a cul-de-sac from which there may be no return, a "coevolutionary shift towards parasitic specialization" or towards mutualistic symbiosis, considered by Luttrell (1974) to represent a "progressive stage of biotrophic parasitism." Clearly, these genera show an advanced physiological specialization and a dependence on the host for complex organic nutrients.

If we suppose that *Conidiobolus*-like Entomophthorales evolved during the Devonian in tropical fern forests, exploiting stressed, and eventually healthy, insects crawling through the forest litter, then adaptation to and colonization of mobile hosts would have involved major changes in spore dispersal. The development of a forcibly discharged asexual spore was necessary for Entomophthorales to contact host insects effectively and separated this order at an early stage from the other Zygomycotina. Further refining of the dispersal mechanism occurred, presumably to reduce the chances of missing the target, with the ability of primary ballistospores to produce secondary spores. The latter, passively dispersed spores, were further modified to maximize the chances of successful contact with passing insects, developing sticky coverings or patches and adhesive appendages ("haptors"). The host range of the primitive Entomophthorales was probably wide, and any insect picking up the spores must have been a potential host. However, as the insects were evolving defence mechanisms to overcome this threat, the fungi were also developing better tactics, becoming more specialized and host specific. With the advent of flying insects, the challenge would have increased immeasurably, resulting in wide generic and species diversity within the entomopathogenic fungi. Ballistospores are ill-adapted for soil survival and are dependent for their production and dispersal upon epigeal habitats, that is, on freely-exposed hosts. Those insects that hid away when infected would easily have suppressed any subsequent spread of inoculum. The failure to develop organized hyphal structures, to lift the sporogenous cells into an aerial niche, must have restricted the range of insects the Entomophthorales could exploit. Many types of insects with entomophthoralean infections die in exposed positions and there are numerous examples of diseased hosts climbing vegetation (Evans, 1982); the syndrome is often referred to as "summit disease". This phenomenon has led to speculation that infected insects alter their behavioural patterns to favour fungal dispersal (King and Humber, 1981). Certainly, any host that sought refuge when infected would present insurmountable difficulties for efficient ballistospore dispersal from sessile sporogenous cells. Did the Entomophthorales succeed in adapting only to those insects whose behavioural patterns ensured efficient dispersal? It could be argued that the fungus produces substances that affect the nervous system of the host by inducing a positively phototrophic response. Conversely, the host could act to reduce the effectiveness of the pathogen by moving away from population centres of healthy individuals. For example, sugarbeet aphids, which typically live and feed below ground, emerge when infected by *Erynia aphidis* and die amongst the vegetation (Harper, 1958). Similarly, *Entomophthora*-infected carrot flies move away from the crop; diseased females change their egg-laying habits and fly upwards to deposit their eggs on the tops of hedgerow trees (Eilenberg, 1986). By such actions, vertical transmission of the pathogen within the population is reduced but, conversely, horizontal transmission between populations would be enhanced. The development of rhizoids, which serve to anchor the host to the substratum, is considered to represent an advanced feature for the Zygomycotina (Madelin, 1966). These are specialized structures that ensure that the sporulating cadaver remains an effective source of aerial inoculum for as long as possible.

In the majority of entomophthoralean infections, the defence alarm systems of the insect detect the presence of the intruder at an early stage, and appropriate action is taken to limit the spread of disease. In the purportedly more highly evolved genera, the infection apparently goes undetected until a late stage. As discussed previously, mycelium of *Strongwellsea* can reach the host brain without affecting host behaviour. Similarly, *Massospora* species, although causing major tissue damage in the abdomen of the cicada host, fail to trigger the central nervous system. Indeed, the dispersal of the asexual spores of these fungi, which are liberated from the disintegrating abdomen of active hosts, is dependent upon the host remaining alive and behaving normally. Infected male cicadas attempt to attract and copulate with females, and even continue to feed (Soper, 1963). Speare (1921) already appreciated the evolutionarily advanced condition of the genus *Massospora*, noting that the firing of ballistospores from dead hosts was extremely inefficient, "... compared with the process which takes place in the present instance, in which the live, actively moving infected host mingles promiscuously with its fellows". He investigated the remarkable relationship between *M. cicadina* and the periodic cicada (*Magicicada septendecem*) which hibernates in the soil for 17 years. The fungus has perfected a resting spore that survives in the soil and germinates when the adults emerge from the pupal cases. There is, therefore, an amazing synchrony between spore germination and insect pupation that ensures that infection occurs during this ephemeral period in the soil when the insect is susceptible. Young, infected cicadas produce asexual spores that are transmitted from the decaying abdomen as the insects move within the population. Older insects that become infected, instead of releasing spores, remain intact and return to die in the soil, the abdomen becoming filled with resting spores. The cycle renews 17 years later. This bizarre example of insect dormancy could be in part the result of a long period of evolution with entomopathogenic fungi, a method of escaping fungal attack. If so, the pathogen has succeeded in staying with its only host by developing an equally complex and highly synchronized life cycle.

Certain aquatic Entomophthorales have evolved equally sophisticated life cycles to match the behavioural patterns of their hosts. It has been shown recently that a group of *Erynia* taxa, attacking aquatic insects belonging to the Diptera, Plecoptera, and Trichoptera, produce up to four spore types to facilitate dispersal, both aerially and aquatically, revealing unusual plasticity in spore morphology and mode of germination (Descals *et al.*, 1981; Descals and Webster, 1984). In contrast to some of the *Entomophthora* examples detailed above, *Erynia*-infected female simulid flies do not appear to change their behavioural patterns and die in well-defined oviposition sites, on the undersides of rocks, firmly attached by rhizoids. Globose asexual spores are produced around sunset, the peak hours of oviposition, maximizing the chances of contact with flying females. Up to 10% of the sampled healthy insects were found to be carrying infective spores (Hywel-Jones, 1986). A possible avoidance mechanism by the insect may have been to increase the numbers of breeding sites in order to keep one step ahead of the pathogen, necessitating a division of labour in its spore types to reach scattered host populations along the fast-flowing streams and rivers that the Simuliidae inhabit.

It is also of interest to consider the origins of the only non-entomophthoralean representative of the entomopathogenic Zygomycotina, *Sporodiniella umbellata*. This mucoraceous pathogen causes epizootics amongst homopteran hosts in some tropical tree crops (Evans and Samson, 1977). It is pertinent to speculate how this seemingly anomalous fungus evolved within the essentially saprophytic Mucorales and adapted to living insects. Some related genera are mycoparasites, probably with a chitin-degrading enzyme system, and may have provided the impetus for the transition from general saprobe to exploiter of chitinous insect remains, then to weakened, and finally to healthy insects. The fungus has also been collected on insect debris and hence should be considered as a facultative rather than an obligate parasite, in sharp contrast to most of the entomopathogenic Zygomycotina. Why more of the mucoraceous fungi did not adopt an entomogenous habit is obscure. Possibly the group has so successfully exploited the saprophytic niche that there were no competitive pressures towards parasitism to ensure survival. The unique umbelliform sporogenous structures represent an adaptation to an entomogenous habit. The interlocking sporangiophores may ensure that the spores are projected or directed onto any insect that disturbs the structures, either by direct contact or by wing movements.

Ascomycotina

Amongst the entomopathogenic Ascomycotina, Clavicipitales is the most important order, rivalling the Entomophthorales in complexity and diversity "... because they probably represent the end products of an ancient coevolution with arthropods ..." (Samson *et al.*, 1988). All entomopathogenic species of the Ascomycotina can be classified as hemibiotrophs (Luttrell, 1974). They possess a well-developed parasitic phase within the haemocoele of the host, which sooner or later induces host death, allowing the fungus to colonize the cadaver saprophytically and subsequently sporulate on it. Pathogenesis, therefore, is considerably different from that in the Zygomycotina, or at least in the Entomophthorales, where typically the pathogen dies, or ceases growth, when the host dies. Death of insects invaded by Ascomycotina is invariably due to toxins released by the fungus (Roberts, 1981). As discussed earlier, some primitive Entomophthorales also produce toxins, but rapid host death is not evolutionarily advantageous in the absence of mechanisms to inhibit competing saprophytes. It would appear that the Ascomycotina successfully overcame this hurdle by developing secondary metabolites with antibiotic activities, thereby removing any pressure to maintain the parasitic phase longer than necessary. Toxins probably build up in the haemocoele as the yeast-like parasitic cells colonize the circulatory system. There is no evidence that true mycelium is ever produced inside the body cavity of the living host, in sharp contrast with the Entomophthorales. There is no doubt that the entomopathogenic Ascomycotina evolved at an early stage in moist tropical habitats, as evidenced by their great diversity in tropical ecosystems, particularly rainforests, compared with the relative paucity of

Entomophthorales (Evans, 1982). They appear to have exploited most orders of insects, and in tropical forest situations they exert a significant and continual control of insect populations, especially of Hymenoptera and Homoptera. In the case of ants, disease appears to be maintained at a constant or enzootic level partly by the activities of the infected hosts (Evans and Samson, 1982, 1984). Diseased ants avoid normal haunts (ant trails, nests), radically alter their behavioural patterns, and die in selected niches. Avoidance of nests is, of course, particularly relevant in social insects and this may be a method by which the highly organized ants reduce the chances of an infectious disease reaching the colony. Many ground-dwelling ponerine ants, when infected by *Cordyceps*, climb vegetation and die in exposed positions, grasping the substratum with legs and mandibles, analogously to the summit disease syndrome in Entomophthorales; the arguments above (p. 157) may apply. Conversely, *Cordyceps*-infected arboreal ant species often descend to hide in tree bark and leaf litter. *Cordyceps* species are well-adapted to hosts that are buried or otherwise hidden, possessing highly coordinated structures that are positively phototrophic and capable of moving considerable distances through soil and litter, and over vegetation. A range of spore types may be produced to increase the short- and long-term possibilities of reaching the insect target. Ascospores of *Cordyceps* are forcibly discharged, and in more advanced groups they fragment on release, from perithecioid ascomata with asynchronous maturation; this is an important strategy where an elusive or low density host is involved. Two or three distinct anamorphs may also be present; usually a dry-spored, air-dispersed anamorph; a short-lived, rain-splashed, slime-spored anamorph; and a durable, mucilaginous anamorph. All may occur on phototrophic synnemata of varying complexities. Those of the latter anamorph often radiate laterally from the host onto the substratum before producing vertical fertile heads that persist after the host has disappeared. The resources of the fungus thus ensure a steady return (enzootic) rather than placing all on the chance of a jackpot (epizootic). Balanced tropical forest ecosystems do not favour the latter.

Coccids with ascomycete infections are especially prominent in tropical forests (Evans, 1974, 1982): the brightly-coloured stromata are conspicuous against fallen leaves, and led Petch (1925) to wonder how these homopterans survived in such habitats. *Hypocrella* and *Torrubiella* species show a predilection for coccid hosts, but the teleomorphs are less commonly encountered than their respective *Aschersonia* and *Verticillium* anamorphs. These are slime-spored and superbly adapted for water-borne movement over leaf surfaces. Such is the efficiency of this short-distance dispersal that once a colony becomes infected it is virtually impossible to find healthy individuals. How did the coccids, therefore, adapt to the threat from the entomopathogenic Ascomycotina and achieve ecological stability? Harlan's (1976) views on plant disease are pertinent here: "Gene complexes that permit the host to avoid epidemics have an advantage over those that allow a fraction of the host population to survive them"; and "The forests of the wet tropics . . . are immensely rich in species deployed it seems for maximum protection against disease epidemics." Thus, it was by avoidance, adaptation to, and speciation on the diverse tree hosts in tropical forests, spatially separating colonies of the same coccid taxon. Diversification of the flora in the pre- and post-tropical forests led to increased specialization of the plant-sucking homopterans and, of course, coevolving host specialization in the parasites. Successful horizontal dispersal of the pathogens between coccid colonies would have decreased proportionally with increasing adaptation to and restriction of their hosts to certain trees. The dry ascospores would then have functioned for long-distance dispersal. Although phototrophic aerial structures were not necessary, because of the exposed habit of the host, ballistospores remained essential for efficient movement from the still laminar layers surrounding the leaves into air currents. In the entomogenous ascomycetes inhabiting tropical forests, there is always a delicate balance in division of resources between production of the vertically-dispersed, slime-spored anamorph (*Aschersonia*) and the horizontally-dispersed, dry-spored teleomorph (*Hypocrella*). This group of scale-insect pathogens is taxonomically diverse (Petch, 1921); the degree of specialization is speculative since few of the insect hosts have been identified (Evans, 1982). However, there is an apparent restriction to soft scales (Lecaniidae) and Aleyrodiidae. The armoured scales (Diaspidiidae) have quite separate pathogens, mainly in the Hypocreales, which lends support to the theory of Steinhaus (1949) that primitive scale insects were not parasitized by a common ancestor but that these Ascomycotina evolved after the coccid subfamilies were delimited. Significantly, *Hypocrella* on Lecaniidae are morphologically distinct from those on Aleyrodiidae. My observations in tropical forests suggest that Lecaniidae were probably the dominant plant-sucking homopterans in ancient tropical ecosystems and that the Clavicipitales were the dominant entomopathogenic group. Adaptation to Aleyrodiidae came later, as did the evolution of the entomopathogenic Hypocreales, possibly when the insects radiated from tropical forests. Certainly, hypocrealean entomopathogens are rarely encountered in tropical forest ecosystems but are common in both tropical and subtropical tree crops (disturbed) habitats, such as cocoa, citrus and pine plantations.

Hyphomycete anamorphs

The entomopathogenic hyphomycetes all probably belong to Ascomycotina and share a similar pathogenesis. Only genera for which the teleomorph is

Figure 1. **A**, Cocoa weevil (*Pantorhytes plutus*) infected by *Beauveria bassiana* in Papua New Guinea; the dying insects fall to the ground from their normal habitat in the tree canopy (Photograph C. Prior; × 3). **B**, Coleopteran larva, originally buried in log, with *Cordyceps* stroma emerging from the insect tunnel and forming brightly-coloured fertile heads with immersed perithecia, Brazil (× 1). **C**, Dipteran infected by *Erynia* sp. (Entomophthorales), clinging to and attached by fungal rhizoids to rice leaves, Indonesia (× 8). **D**, Moth, freely exposed on and fastened to a tree branch by fungal mycelium, with *Cordyceps* perithecia formed superficially on the dorsal surface and cylindrical *Akanthomyces* synnemata, Ecuador (× 1).

completely unknown are therefore discussed here. Two of the commonest genera, *Beauveria* and *Metarhizium*, have been well researched because of their potential for biocontrol. In tropical forests, however, these genera are rarely encountered on insects, although *Beauveria* can be found colonizing insect remains in the soil. These genera may have a well-adapted soil survival phase, either as spores or saprophytic mycelium, differing from Clavicipitales anamorphs in this respect. Biotypes of *Metarhizium* have occasionally been collected on soil-inhabiting insects in tropical forests. The structure of the synnemata is highly organized and more complex than that of specimens in non-forest habitats. These fungi appear to have successfully adapted to a wide range of insect hosts outside of the forest ecosystem. As shown earlier, *Beauveria* has been around since at least the Cenozoic, when most extant insect families were arising in a variety of ecosystems. Many of the Ascomycotina that made the transition from buffered tropical forest habitats to more seasonal, less predictable situations, appear to have sacrificed their teleomorph, and thus the ability for genetic recombination, at an early stage in exchange for the massive production of dry conidia. This may be correlated with seasonal population explosions of host insects, the fungus surviving in the soil during periods of low host density or unfavourable climatic conditions. Thus, these fungi were ideally equipped to invade disturbed habitats created by man's activities.

Apart from resistance mechanisms, what methods have potential hosts adopted to minimize or escape disease outbreaks? There is evidence that *Beauveria*-infected, arboreal lepidopteran larvae and weevils fall to the ground (Samson and Evans, 1982; Prior *et al.*, 1988), essentially removing themselves from the healthy population. This contrasts with the clinging habit described for previous insect-pathogen associations. Another strategy to combat pathogens with good survival ability may have been to change breeding habits. For example, *Oryctes* beetles in south-east Asia are attacked by a highly specialized *Metarhizium* strain that, once inside the breeding site, is lethal. The beetle may have avoided such catastrophic encounters by decreasing the size of egg clutches whilst correspondingly increasing the number of egg-laying sites.

Entomogenous mutualistic symbionts

The relationship between fungus and insect is mutualistic when both partners benefit from the association. There is no evidence that such an interdependent relationship, the climax of coevolution, has been achieved in either the entomogenous Zygomycotina or Ascomycotina. However, in the Basidiomycotina, Septobasidiales contains nearly 200 species exclusively and mutualistically associated with scale-insect hosts. Most of our knowledge on this highly specialized but basically neglected group we owe to Couch (1937, 1938). He clearly showed that in *Septobasidium* species the fungus provides protection from predators, parasites, and pathogens; and the insect supplies nutrients and a means of dispersal. Nevertheless, the relationship is unbalanced, since the insect can survive without the fungus but not vice versa, although "Such insects are comparatively rare and their life exceedingly precarious" (Couch, 1938). The fungus forms annual rings of mycelium over and around the scale-insect colonies. The covering is extremely tough and varies from 150–1000 μm in depth, deterring predators and virtually eliminating parasitism by hymenopterans, since the ovipositor (200–300 μm in length) can only penetrate the colony edge.

The Septobasidiales form complex coiled haustoria within the host haemocoele, after entering the insect mainly through natural apertures. Nutrients pass from the haustoria via fine penetrating hyphae into the fungal mat. The parasitized scale-insects, which live on shrubs and trees, continue to feed on the plant host, thereby ensuring a continuous flow of nutrients as long as the insect remains alive. Couch considered that such individuals usually live as long as, if not longer than, non-parasitized scales. However, parasitized insects fail to mature normally, becoming dwarfed and sterile. The amount of fungal growth and, therefore, the size of the colony is directly dependent upon the number of parasitized scales, but the fungus colonizes only a proportion of the scales whilst creating a stable microhabitat for progressive generations of insects. Thus, there is a compromise, the fungus opting for a slow-growing but long-lasting colony. The plant host is the only partner adversely affected in this highly developed association. Further evidence for the advanced condition of this relationship, indicating coevolution over a considerable period of time, has been gathered by Couch (1938). He deduced that the scales are chemotactically attracted to the fungus and that, reciprocally, the insects excrete substances that markedly affect fungal growth. The elaborate tunnels and chambers, corresponding to the shape of the particular host insect, that are formed in fungal colonies on trees are never seen *in vitro*, suggesting that the insects actively produce growth modifiers. The basidiospores of *Septobasidium* are disseminated inwards into the colony and long-distance dispersal is effected by crawlers during periodic migration within and between tree hosts.

The nature of the association between species of the much smaller genus *Uredinella* and their scale hosts is more difficult to interpret in terms of mutualism. These fungi have similarly sophisticated haustoria within the host, but each fungal colony parasitizes but a single scale-insect. The advantages that accrue to the insect are therefore dubious and the relationship is probably one of unilateral parasitism. Direct mortality is not ascribed to the fungus, since it is beneficial to the latter to cause minimal disruption to the host. Death of the insect results in cessation of fungal growth and subsequent senescence. The parasitized scales are almost certainly incapable of reproduction. Consequently, although the fungus affords temporary protection to individual scales, it has a negative effect on insect populations. In coevolutionary terms, members of the genus *Septobasidium* have approached the ultimate goal with their hosts, that of mutualistic symbiosis. The genus *Uredinella* has diverged from this path and has perhaps laid the foundations for other groups of parasitic Basidiomycotina, as discussed below.

Entomogenous ectoparasites

The group of entomogenous ectoparasites is represented almost entirely by the Laboulbeniales (Ascomycotina), which are exclusively entomogenous. There are over 1800 described species in more than 100 genera (Hulden, 1983), and many more await discovery. Thaxter devoted much of his considerable talent to describing these highly specialized parasites in a series of papers from 1896 to 1931. Steinhaus (1946, 1949) summarized and popularized Thaxter's contributions, and recently Tavares (1985) has added significantly to our understanding of this group.

Although they colonize the cuticle of insects, deriving all their nutrients from the host, the Laboulbeniales cause little damage and are considered to be true commensals (Steinhaus, 1949; Hulden, 1983). They are obligate symbionts and cannot survive without the host. Their vegetative structure is much reduced and limited to a foot region functioning as a holdfast and haustorium. However, the Laboulbeniales have developed a unique ascus structure that suggests separation at an early stage from other Ascomycotina. This is reinforced not only by extremes of specialization between insect species, but also by specialization within the same host. Each fungal taxon appears to be limited to a small group or genus of insects, usually in the Coleoptera and Diptera, and to a specific area of colonization on the host body. Some species may occur only in one position, on one sex of a single insect species. Since transmission of the fungus is effected predominantly during copulation, the parasite has adapted not just to the copulatory behaviour of the host but also to the region of the body where maximum transmission is ensured. Tavares (1985) hypothesized that the precursors of the Laboulbeniales parasitized marine arthropods and could have colonized the primitive terrestrial descendants of the true beetles during the Carboniferous (*c.* 300 million years ago). As the Coleoptera evolved during the Triassic and Jurassic and the modern beetle families were delimited, the major groups of Laboulbeniales presumably coevolved. However, the relationships of this group are still uncertain (M. Blackwell and D. Malloch, personal communication).

Several hyphomycete genera are ectoparasites on termites (*Termitaria*, *Mattirolella*). In common with the Laboulbeniales, they form haustoria within the insect cuticle but do not penetrate beyond (Khan and Kimbrough, 1974a). Observations suggest that fungal-infected termites do not behave abnormally and that host damage is minimal (Khan and Kimbrough, 1974b). They are therefore regarded as commensals. Their unique morphology and host specificity indicates a similar period of coevolution between fungus and insect as proposed for the Laboulbeniales, perhaps from a common ancestral stock.

Entomogenous endoparasites

The Trichomycetes (Zygomycotina) are obligately associated with the cuticular surfaces of insects, and of arthropods in general. But, in contrast to the ectoparasites, they are almost entirely confined to internal integuments within the digestive tract. They have been poorly studied, but most evidence indicates that they have developed a commensalistic association with their hosts. Moss (1979) concluded that the Trichomycetes were basically nonparasitic, since they mainly occupy the hindgut where only limited nutrient uptake by the host occurs. This association, therefore, has not resulted in any pressures on the host to modify its morphology, physiology, or behaviour. In other words, coevolution has not occurred in the relationship. Finally, he hypothesized that the Trichomycetes evolved from free-living aquatic fungi that originally exploited an oligotrophic niche. However, at least one species invades the ovaries of Simuliidae (Diptera) with a subsequent reduction in egg production (Moss, 1986). Their nutritional status is uncertain, as is their occurrence within and specialization between insect families.

GENERAL THOUGHTS AND FURTHER SPECULATIONS

The present evidence indicates that fungal parasitism of insects arose *de novo* in the phylogenetically distinct subdivisions of the fungi.

Whether the terrestrial fungi developed from marine saprophytes or parasites has been the subject of much debate. Savile (1955, 1971) considered that the saprophytic habit followed from a parasitic existence: "How else, indeed, could filamentous fungi have left the aquatic habitat except shielded by the tissues of their hosts?" Another school of thought takes the opposite viewpoint (Leppik, 1965; Luttrell, 1974). I favour the hypothesis of Cain (1972), that the precursors of extant fungi, at least in the Ascomycotina and Basidiomycotina, were autotrophic and that saprotrophy evolved from such beginnings with subsequent adaptation to and parasitism of land plants and insects, or, in some instances, a direct transition from autotrophy to parasitism. Cain speculated that ancestral Basidiomycotina and Ascomycotina arose independently and adapted to moist tropical lands as autotrophs from a marine environment. He termed these "basidiophytes" and "ascophytes" respectively.

If we consider the basidiophytes in relation to Septobasidiales, certain basidiophytes must have become epiphytic early in vascular plant evolution. The holdfast rhizoids eventually penetrated the "host" plant and obtained nutrients from it. Heterotrophism replaced autotrophism and photosynthetic ability was sacrificed. Could some of the ancestral Septobasidiales, which adapted to an autotrophic, epiphytic existence, have come into intimate and sustained contact with primitive plant-sucking homopterans, perhaps seeking shelter beneath the fungal mycelium? The sequence of host penetration, diversion of nutrients, and loss of chlorophyll would then have followed, involving a direct transition or jump from autotrophic to parasitic nutrition. Consequently, such fungi would never have developed a saprophytic ability and evolution would have favoured a non-lethal association with the host, keeping it alive as long as possible. *Uredinella* has not achieved this balance, having constantly to switch from host to host, whereas *Septobasidium* has perfected a mutualistic relationship and is a much more successful parasite of scale-insects, if this is measured in terms of ubiquity. *Septobasidium* appears to be at the end of an evolutionary line. *Uredinella* has not reached this point but may have escaped the dead-end. Could it have invaded the plant host via the insect stylet, effectively by-passing the intermediate nutrient sink of the insect? In fact, Couch (1937) compared the sporogenous structures of the genus with those of primitive rusts. Leppik (1965) vigorously supported the theory that the rusts evolved from within this group, regarding the basidia of *Uredinella* as an intermediate step in the delimitation of rust teliospores. The route suggested above, towards the evolution of a biotrophic parasitism of vascular plants, would have been irreversible. This transition from an autotrophic to a heterotrophic mode of nutrition in an epiphytic habitat favoured parasitism and not saprotrophy. The possibility of a pre-insect association in rust evolution is intriguing.

According to Cain, the ascophytes were also evolving in these wet, tropical soils. Almost certainly, the two main entomogenous Ascomycotina groups, entomopathogens and ectoparasites, separated at a very early stage. The Laboulbeniales may even have "jumped" from a marine arthropod directly to a primitive terrestrial insect, thus maintaining a continuous parasitic habit. It would seem that ectoparasites never went through a saprophytic phase, following a parasitic line throughout and ending in a specialist refugium from which there was no return and no advancement. It is difficult to envisage how they could achieve a mutualistic relationship. An equilibrium has been reached, with the insect losing little ground, and the parasite receiving just enough to survive.

The entomopathogenic Ascomycotina were much more aggressive. They entered a parasitic refuge from a saprophytic free-for-all, presumably to escape from competitors, to exploit the increasingly successful and abundant insects. But, by evolving protective mechanisms (antibiotics), they were able to hold on to their saprophytic habit and benefit from both lifestyles. There were no pressures to proceed either to obligate parasitism or to mutualism. The fungus could not increase in fitness or perhaps decrease in fitness in such restrictive relationships.

The entomogenous Zygomycotina and fungus-like Mastigomycotina may have evolved as parasites on marine arthropods or as saprophytes on their remains. Saprotrophy in the Zygomycotina, as was previously discussed, appears to be of ancient origin. The Mucorales adopted almost exclusively this mode of nutrition; *Sporodiniella* is the exception, whilst the Entomophthorales gradually evolved towards insect parasitism. Failure of the Entomophthorales to elaborate antimicrobial defences has resulted in obligate parasitism. They entered the parasitic refugium and the one-way system with a premature ending. It is difficult to envisage any continuation towards mutualism, because these fungi would appear not to be equipped to offer anything in return.

I wish to conclude with further speculations on the entomopathogenic fungi. Fungi and insects evolved in moist, tropical environments; the extant tropical forests are rich in entomopathogens, particularly Ascomycotina, and their anamorphs. The majority of these fungi are strikingly coloured. They advertise themselves in the leaf litter, in logs and tree trunks and on branches and leaves. This signal-type coloration can be expected to be adaptive in the same way as that of coloured flowers of Angiosperms. Does it serve to draw in foraging insects and, if so, is it selective? For example, does a bright orange *Cordyceps* stroma arising from an elaterid beetle buried in a log attract passing Elateridae? Preliminary observations on immature *Cordyceps* species maintained in the laboratory indicate that, in addition to bright pigmentation, the stromata also exude nectar-like substances as they develop. This is especially characteristic of the slime-spored anamorphs. If insects are attracted by this combination of colour and viscid secretions, few would be potential hosts of such specialized pathogens. This suggests a random vector dispersal of the sticky spores, but with no accompanying mechanism for transmission. However, the spores may remain on, or superficially within, the exoskeleton until death of the vector and then exploit this niche to produce a dry-spored anamorph (the ancestral soil-litter inhabitant). This would greatly increase the range of the fungus whilst improving its chances of reaching a low-density host target. Similarly, the ascospores may be insect-dispersed as well as air-dispersed. Ascospores are released violently by external stimuli such as changing light intensity. Wing movements of attracted insects may provide an alternative stimulus.

In conclusion, we can still only offer speculation to explain much of the biology of tropical entomopathogenic fungi because they have been so little studied. They may, however, have played a larger role in insect evolution than is currently acknowledged. Harlan (1976) suggested that diseases may have been important in determining species diversity in tropical forests: "... diseases are likely to have had the most impact on host evolution in the warm wet parts of the world". It is to natural ecosystems, such as tropical forests, therefore, that one must look for evidence of coevolution in such groups.

REFERENCES

Ben-Ze'ev, I. S., Keller, S. and Ewen, A. B. (1985). *Entomophthora erupta* and *Entomophthora helvetica* sp. nov. (Zygomycotina:Entomophthorales), two pathogens of Miridae (Heteroptera) distinguished by pathobiological and nuclear features. *Can. J. Bot.* **63,** 1469–1475.

Boucias, D. G. and Latge, J. P. (1986). Adhesion of entomopathogenic fungi on their host cuticle. In "Fundamental and Applied Aspects of Invertebrate Pathology" (R. A. Samson, J. M. Vlak, and D. Peters, eds), pp. 432–434. Foundation of the Fourth International Colloquium of Invertebrate Pathology, Wageningen.

Cain, R. (1972). Evolution of the fungi. *Mycologia* **64,** 1–14.

Charnley, A. K. (1984). Physiological aspects of destructive pathogenesis in insects by fungi: a speculative review. In "Invertebrate–Microbial Interactions" (J. M. Anderson, A. D. M. Rayner, and D. H. Walton, eds), pp. 229–270. Cambridge University Press, Cambridge.

Cooke, M. C. (1892). "Vegetable Wasps and Plant Worms". Society for Promoting Christian Knowledge, London.

Cooke, R. (1977). "The Biology of Symbiotic Fungi". Wiley, London.

Couch, J. N. (1937). A new fungus intermediate between the rusts and *Septobasidium*. *Mycologia* **29,** 665–673.

Couch, J. N. (1938). "The Genus *Septobasidium*". University of North Carolina Press.

De Bary, A. (1879). "Die Erscheinung der Symbiose". Trubner, Strasbourg.

Descals, E. and Webster, J. (1984). Branched aquatic conidia in *Erynia* and *Entomophthora sensu lato*. *Trans. Br. mycol. Soc.* **83,** 669–682.

Descals, E., Webster, J., Ladle, M. and Bass, J. A. B. (1981). Variations in asexual reproduction in species of *Entomophthora* on aquatic insects. *Trans. Br. mycol. Soc.* **77,** 85–102.

Eilenberg, J. (1986). Effect of *Entomophthora muscae* (C.) Fres. on egg-laying behaviour of female carrot flies (*Psila rosae* F.). In "Fundamental and Applied Aspects of Invertebrate Pathology" (R. A. Samson, J. M. Vlak, and D. Peters, eds), p. 235. Foundation of the Fourth International Colloquium of Invertebrate Pathology, Wageningen.

Eilenberg, J., Bresciani, J. and Latge, J. P. (1986). Primary spore formation and discharge in the genus *Entomophthora*. In "Fundamental and Applied Aspects of Invertebrate Pathology" (R. A. Samson, J. M. Vlak, and D. Peters, eds), pp. 182–185. Foundation of the Fourth International Colloquium of Invertebrate Pathology, Wageningen.

8 Metabolites in the Coevolution of Fungal Chemical Defence Systems

D. T. WICKLOW

Northern Regional Research Center, Agricultural Research Service, United States Department of Agriculture, Illinois, USA

Abstract

This chapter considers the ecology and evolutionary biology of fungal chemical defence systems. Fungal toxins are considered analogous to chemical defences of higher plants in detering potential predators. Predation as a selective force has shaped the chemical defense systems of fungi. Toxins simultaneously produced by the same fungus can have synergistic effects. Fusaric acid, from pathogenic *Fusarium* species, can synergize the toxicity of phytotoxins produced by plants that the fungus infects. Evidence is offered that fungal chemical defences are distributed in direct proportion to the risk of the particular tissue and the value of that tissue in terms of fitness loss to the organism as the result of an attack on that tissue. Fungal sclerotia contain toxic metabolites not produced by fungal hyphae. The survival of sclerotia is examined in connection with the means by which these structures deter potential mycophagists (e.g., arthropods, rodents).

Detritivorous insects are better able to tolerate moulded and mycotoxin-contaminated resources than are herbivorous insects. Selective mycophagy can have an indirect effect on the outcome of competitive encounters between fungal species and thus influence processes of decomposition in nature. Compensatory growth or reproductive responses of fungi to moderate grazing suggest that low arthropod-grazing intensities can optimize an organism's competitive fitness. No one has attempted to demonstrate that insects sequester fungal toxins to defend themselves against predators, use fungal toxins as aids in the recognition of their hosts, or benefit nutritionally from the presence of nitrogen-rich toxins.

Understanding the relationships between the antagonistic or toxigenic properties of fungi and their ecological status can contribute to the development of effective strategies for biological control, and to the search for fungal metabolites active against competitors and pests. For example, some seeds may protect themselves from predators by harbouring fungi that manufacture toxic metabolites.

COEVOLUTION OF FUNGI
WITH PLANTS AND ANIMALS
ISBN 0-12-557365-0

associated with feeds and food, and the mycotoxins they produce is provided by Frisvad (1986). Much of this information has been obtained over the past two decades. It is unfortunate that neither this chemical database, nor data on the mycotoxin profiles of individual fungal strains, was available to Sinha and his colleagues when they initiated ecological studies on the feeding behaviour and biology of stored product arthropods from different feeding guilds (Sinha, 1964, 1966; Loschiavo and Sinha, 1966; Sinha *et al.*, 1969; Sinha, 1971); these accounts contain suggestions that feeding and oviposition preferences may be related to the presence or absence of toxic fungal metabolites. For example, in choice tests involving different stored product arthropods, toxigenic species of *Aspergillus* and *Penicillium* appeared consistently to be least favoured. Differences in the reproductive success of arthropods tested before and after a phenotypic change in the stock culture (Sinha, 1966) might be explained by a loss in fungal ability to produce toxic metabolites. Wright *et al.* (1980a,b,c) have shown that beetles' avoidance of specific isolates of *Penicillium purpurogenum*, *P. martensii* and *P. citrinum* is correlated with life history data showing inhibited growth and larval development on diets comprised of these cereal-infesting moulds. Their results also show that strains recently isolated from moulded grain proved more toxic to the beetles than ones that had been repeatedly subcultured any may have lost their ability to produce specific metabolites.

It has been shown repeatedly that herbivorous insects have evolved mechanisms for detoxification of plant defensive substances. Most commonly the defensive substance is metabolized to a less toxic derivative that the insect then excretes (Dowd *et al.*, 1983), but there are no published accounts demonstrating the detoxification of fungal toxins by insects. Dowd and van Middlesworth (unpublished) found that the detritivorous beetle *Carpophilus hemipterus* can readily deacetylate radiolabelled monoacetoxyscirpenol (MAS) produced by certain seed-infesting species of *Fusarium*. The gut tissues of *C. hemipterus* were 8–10 times more effective, on a per milligram of protein basis, in metabolizing MAS than were those of two herbivorous lepidopteran insects (*Heliothis zea* and *Spodoptera frugiperda*).

Jaenike (1985) demonstrated an important selective advantage of amanitin tolerance in mushroom-breeding species of *Drosophila*. Amanitin-tolerant drosophilids are able to breed in mushrooms that are toxic to intestinal parasitic nematodes that infect *Drosophila* and render the adults infertile. Here the physiological/metabolic cost associated with amanitin tolerance is at least equalled by the benefit of being able to breed in these fungi and prevent nematode parasitism.

Some herbivores are known to sequester or store plant toxins for their own defence (Rothschild, 1973). I know of no reports demonstrating that insects sequester fungal toxins to defend themselves against predators. Rawlins

(1984) makes the observation that lepidopteran lichenivores in the Lithosiinae (Arctiidae) are protected by toxins from the lichen thalli that their larvae feed upon. This, he suggests, is reflected in their aposematic coloration as adults, involving bright reds or yellows, metallic blues, and whites. However, no chemical data were obtained to show that lichen acids are indeed sequestered by the moths and, therefore, this hypothesis remains to be tested. Some fungus-feeding beetles that graze primarily upon the surfaces of polypores ("Handsome fungus beetles", Endomychidae; "Pleasing fungus beetles", Erotylidae) are brightly coloured in shades of orange to red. These colours could warn insectivorous birds and other animals that the beetles have sequestered potent fungal toxins.

Rhoades (1985) points out that most herbivores use plant defensive metabolites as aids in the recognition of their hosts. However, herbivores may benefit nutritionally from the presence of nitrogen-rich plant defensive chemicals if they pay the cost of making specific enzymes to degrade such compounds (Rosenthal *et al.*, 1977; Bernays and Woodhead, 1982). Janzen (1978) predicts that the more toxic the food, the more physiologically and biochemically specialized the animal must be to use it, and the more likely it is that such a restricted population will become dependent on the secondary compounds in that food. Although some fungi are known to attract insects (Evans, this volume, Chapter 7), I know of no examples in which toxic fungal metabolites have been shown to be nutritionally beneficial to invertebrate mycophagists. Examples might be sought amongst Phalacrid beetle predators of ergot sclerotia. The ability to overcome the effects of fungal toxins is well demonstrated by the Phalacrid beetle *Acylomus pugetanus*, which feeds on sclerotia of ergot on several grasses (Steiner, 1984). Tunnelling by the beetle larvae can render the sclerotia non-viable. One might expect that this Phalacrid beetle produces a suite of important detoxification enzymes. Here the host specificity of the sclerotium predator fits the model of a small, prey-specific seed predator (Janzen, 1978).

Plants produce inducible defences against herbivores in response to herbivore damage or simulated herbivore damage (Rhoades, 1985). Two fungal responses to injury may be shown to be inducible defences: the release of milky latex upon injury by basidiomata of *Lactarius* is analogous to the situation in plants that release latex upon being injured to "gum up" the mouth parts of insect herbivores; and the instantaneous tissue-colour changes brought about by enzyme-catalysed oxidation reactions following insect wounding in Boletales (Bruns, 1984). Such compounds (e.g., derivatives of pulvinic acid, grevillins, diphenylcyclopentenones, and benzoquinones) could represent a rapid fungal defensive response to mycophagists, resulting in decreased nutritional quality of the fungal tissue.

SELECTION FOR TOXIN-PRODUCING ABILITY

While there is no direct experimental evidence of selection for ability to produce mycotoxins among naturally occurring fungal populations, there is some indirect evidence.

(1) Tropical plants are known to produce kinds and quantities of toxic alkaloids in what is regarded as an evolutionary response to a larger number of potential predators and parasites in the tropics. Moving towards the equator, the proportion of alkaloid-bearing plants increases, the mean content of alkaloids in leaves increases, and the toxicity of the alkaloids increases (Levin, 1978). The fungal toxin cyclopiazonic acid was produced primarily by strains of *Aspergillus flavus* isolated from warmer latitudes (Wicklow and Cole, 1982). According to Floss (1976), the structure of cyclopiazonic acid suggests a biosynthetic origin very similar to that of the ergot alkaloids.

(2) A study of the geographical distribution of aflatoxin-producing yellow-green aspergilli in south-east Asia and Japan revealed that *A. flavus* was rarely detected in soil samples from the northern part of Japan, but was recorded more frequently from the southern region of Japan and most frequently in the Philippines, Indonesia, and Thailand (Manabe and Tsuruta, 1978). In the southern region of the Honshu and Kyushu district, the non-aflatoxigenic strains outnumbered the toxigenic ones. However, in the more southern islands of Japan as well as in the Philippines, Thailand, and Indonesia, the number of aflatoxin-producing strains isolated was substantially greater. Ability to produce aflatoxin could be a response to greater numbers of potential arthropod predators in warmer latitudes. Similarly, in *Stachybotrys alternans*, toxic and non-toxic strains are reported to prevail over the southern half of Europe and non-toxic strains to predominate in the northern half (Forgacs, 1972).

(3) Toxin-producing ability can be lost following repeated subculture. Domesticated strains (koji moulds) of yellow-green aspergilli (*Aspergillus flavus* group; Raper and Fennell, 1965) differ both morphologically and biochemically from wild strains, and Wicklow (1984) illustrates how selection in nature and in the koji environment has produced such fungal adaptations: koji strains were derived from wild and potentially toxigenic strains. Upon repeated transfer of inoculum, the koji strains lost their ability to produce toxic metabolites because in a koji environment these metabolites were no longer of adaptive value to the fungus. Koji moulds are not subjected to the negative effects of predation, since the survival of progeny is man-managed. Loss of mycotoxin-producing ability commonly occurs among strains cultured in the laboratory. For example, Dingley *et al.* (1962) found that production of sporidesmin decreased to about 30% of the original amount after *Pithomyces chartarum* was subcultured about eight times. Kobel and Sanglier (1978) re-

ported that in some ergotoxine-alkaloid-producing strains of *Claviceps purpurea* a decrease in the alkaloid-production capacity is correlated with the loss of purple pigmented colonies. After the first transfer by means of spore suspension, 97% of the colonies were purple; after seven successive transfers, the number of purple colonies had dropped to 22% and alkaloid productivity was 3% of that obtained following the initial transfer.

(4) Bu'Lock (1980) observed that the most potent mycotoxins tend to occur towards the end of a long biosynthetic route in which substances of lower toxicity are intermediates. This suggests that selection has guided the evolution of such biosynthetic pathways. In the case of the biosynthetic pathway leading to aflatoxin B1, Bennett (1983) argues that if toxic polyketides provide a selective advantage against mycophagous arthropods, the evolution of detoxification systems by the arthropods has led to the synthesis of increasingly potent polyketides. This hypothesis is supported by the data of Jarvis *et al.* (1984), who examined the relative toxicity of aflatoxin G1 and selected biosynthetic precursors (sterigmatocystin, versicolorin A, averufin, and norsolorinic acid) to newly hatched, second-, and fourth-instar larvae of the European corn borer *Ostrinia nubilalis*. Newly hatched larvae were affected by aflatoxin G1 levels (32 p.p.b.) that had no significant effect on second-instar larvae, whereas second-instar larvae were affected by aflatoxin G1 levels (250 p.p.b.) that had no significant effect on fourth-instar larvae. Sterigmatocystin at 60 000 p.p.b. was somewhat toxic to newly hatched, second- and fourth-instar larvae, while versicolorin A, averufin, and norsolorinic acid at 60 000 p.p.b. had no significant effect on the larvae.

FUNGAL TOXINS IN INSECT PATHOLOGY

It is clear that some fungal pathogens of insects kill their hosts with toxins (Tamura and Takahashi, 1971; Roberts, 1966). In later stages of fungal infection, insects may exhibit sluggishness, decreased irritability, inability to recover from an inverted position, and partial or general paralysis (Madelin, 1963). These symptoms are not dissimilar from those produced by synthetic neurotoxic insecticides (Charnley, 1984). Wright *et al.* (1982) suggested that, as fungal metabolites that are not effective antibiotics or mammalian toxins have seldom been studied in insects, it might be possible to find fungal products that are potent against insects but that have a low toxicity to mammals. Examples are the destruxins of *Metarhizium anisopliae* and beauvericin from *Beauveria bassiana*; these are toxic to insects but have not been implicated in mammalian mycotoxicoses.

Bennett (1981, p. 413) suggests that the function of aflatoxins is to kill insect vectors of *Aspergillus flavus* "after the fungus is delivered to a food supply"; the insect then "serves as a substrate which the fungus uses to create a large inoculum". While fungal pathogens of insects often sporulate heavily on their dead host, it is not clear how the fungus "knows" when it has arrived at the "food supply" and should begin to make aflatoxins. Moreover, insect vectors of *A. flavus* in agricultural systems already inhabit the field sites of their plant hosts and can successfully deliver an infective inoculum without loss of life. Lillehoj (1981) argues that aflatoxin-producing fungi have formed an association with certain insects that is "symbiotic" in a stable environment but pathogenic in an unstable one. The nature of the mutualistic interaction was not explained, nor did Lillehoj address the evolutionary implications of such interactions in terms of population dynamics. He proposed that aflatoxin (a) provides an advantage to the fungus in competition with other microbes in the insect digestive tract, and (b) functions as a mutagen providing an opportunity for the initiation of genetic plasticity among organisms that could be advantageous in unstable environments. Aflatoxin is a compound having very little demonstrated antimicrobial activity (Burmeister and Hesseltine, 1966; Arai *et al.*, 1967). The advantage gained by a fungus in producing a mutagen that increases the mutation rates of other organisms in its environment is unclear. Ciegler (1983) takes issue with Lillehoj (1981), observing that intrinsic rates of mutation and natural variation in populations already provide genetic plasticity; he also points out that in *Drosophila* there are many more chromosome inversions in complex habitats than in simple ones (Jones, 1981). Entomopathogenic fungi clearly have considerable potential as a source of microbial insecticides (Evans, this volume, Chapter 7).

MYCOPHAGY AND FUNGAL COMPETITION FOR SUBSTRATE RESOURCES

Fungal tissue is equal to plant tissue in its nutritional value to insects (Martin, 1979). Selective mycophagy is recognized as a response, in part, to the presence of specific toxic metabolites in fungal tissues. The feeding behaviours of invertebrate mycophagists can influence the proportions of bacteria and fungi colonizing a resource, and can thus affect patterns of microbial succession and the rate of substrate decomposition (Anderson and Ineson, 1984). For example, the number of coexisting fungal species decreases under the impact of larval (sciarid fly) grazing of microbial communities in rabbit dung (Wicklow and Yocom, 1982). Caddis fly larvae (Trichoptera), feeding on decaying leaves in streams, are also able to choose among a mosaic of "patches" differing in fungal composition and stage of decomposition (Arsuffi and Sukerkropp, 1985). It is not known whether the less-preferred fungi produced toxic metabolites that influenced larval choice. Granary arthropods selectively feed on "field fungi" such as *Alternaria*, and may be partially responsible for the dramatic decline in *Alternaria* population densities in grain bulks infested with this beetle (van Bronswijk and Sinha, 1971). In experimental laboratory microcosms, selective grazing by a collembolan, *Onychiurus subtenius*, significantly reduces the colonizing ability of a sterile dark mycelial isolate from *Populus* leaf litter, giving competitive advantage to a potentially toxic hyaline basidiomycete (Parkinson *et al.*, 1979). Selective grazing of *Marasmius androsaceus* by *Onychiurus latus* restricted *M. androsaceus* to the surface few millimetres of Sitka spruce (*Picea sitchensis*) litter, where the density of *O. latus* is lower because of its susceptibility to desiccation (Newell, 1984a,b). *Mycena galopus*, the other dominant basidiomycete at the site, occurred in the F1 horizon. Curiously *M. androsaceus* was found to have a higher competitive ability because of its colonizing ability, yet it was not dominant in the field. The collembolan showed a strong feeding preference for the mycelium of *M. androsaceus* over that of *M. galopus*, and Newell (1984b) concluded that selective grazing altered the outcome of competition between these fungi in favour of *M. galopus* in the F1 horizon. This suggests that *M. galopus* may either be less nutritious than *M. androsaceus* or produces metabolites that deter collembolan grazing of its mycelium. Using choice experiments, Shaw (1985), examined the feeding preferences of field-collected populations of *Onychiurus armatus* for fungal hyphae of mycorrhizal and non-mycorrhizal basidiomycetes isolated from the same pine plantations. He observed that the least-preferred fungi contained toxic compounds, *Hebeloma crustiliniforme* and *Paxillus involutus* being poisonous to man owing to high concentrations of muscarine and other toxins, while *Rhizopogon luteolus* is toxic to slugs. Shaw (1985) demonstrated a hierarchy in feeding choices of Collembola and suggested that toxins may protect hyphae from grazing by *Onychiurus*.

An approach to understanding the ecological implications of selective mycophagy at the community level, and the role that toxins play, is provided by Lawrey (1980). He designed both field and laboratory experiments showing that the lichen-grazing slug *Pallifera varia* preferred certain species in a community of saxiocolous lichens. These preferences were based, in part, on lichen secondary chemistry, with stictic acid and protocetraric acid appearing to function as antiherbivore compounds in observed slug–lichen interactions. Lawrey (1980) notes that although there has been an enormous amount of information published on the taxonomic significance of lichen secondary compounds, the biological role of these compounds is virtually unknown (Hawksworth and Hill, 1984). Lawrey suggests that *Aspicilia gibbosa*, the most palatable lichen species in his study, is able to maintain itself in the community through a faster growth rate and higher colonizing potential. The coevolutionary implications of secondary metabolites in lichens are considered further by Hawksworth (this volume, Chapter 6).

TOXIN SYNERGISM AND GROUP DEFENCE

Do toxins simultaneously produced by the same fungus have synergistic effects? An example of synergism involving co-occurring phytotoxins suggests that the function of one number of the pair is to inhibit the herbivore's detoxification enzymes (Berenbaum and Neal, 1985). Dowd (1988a) demonstrated that kojic acid synergizes the toxicity of aflatoxin B1 to lepidopteran insect pests of crops; both metabolites are produced by *Aspergillus flavus* and co-occur naturally. In the presence of kojic acid (22 p.p.m.), only one-tenth the amount of aflatoxin B1 (0.25 p.p.m.) was needed to produce levels of toxicity/mortality seen for aflatoxin B1 alone at 2.5 p.p.m. Dowd suggests that kojic acid might be acting synergisticly by inhibiting oxidative enzymes likely to be involved in aflatoxin B1 detoxification. Here the fungus may inhibit the detoxification systems of an insect at comparatively low metabolic cost, thus enabling it to use metabolically expensive toxins more efficiently.

Plant pathogens such as *Verticillium albo-atrum* can influence the acceptability or suitability of plant tissues for phytophagous insects (Kingsley *et al.*, 1983). However, the specific chemicals responsible for the increased metabolic cost to the insect of processing fungus-infected plants are unknown. Karban *et al.* (1987) attribute the ability of fungus-infected plants to deter insect feeding to the induction of phytoalexin production, but perhaps fungal metabolites were also involved. Evidence that fusaric acid, from pathogenic *Fusarium* species, can synergize the toxicity of the plant phytotoxins gossypol (a saponin) and 6-methoxy-2-benzoxazolinone to larvae of *Heliothis zea* has been interpreted as a novel mechanism of defence in which the fungus enhances the toxicity of host plant defences at relatively low metabolic cost (Dowd, 1988b). These results were obtained at concentrations of fusaric acid recorded for naturally infected plant tissues. Dowd suggests that fusaric acid, which is known to chelate metal ions, may complex the iron atom present in insect monooxygenases that are normally responsible for detoxifying the phytotoxins.

Does the potential for group defence exist among assemblages of seed-infesting toxigenic fungi? Janzen (1977) observes that when a number of species are involved in producing the traits of the rotting fruit, they should all be evolutionarily interested in not being eaten and, therefore, are the product of past "group" selection. We might begin to examine how mycotoxin producing ability is distributed among co-occurring populations of seed-infesting fungi. For example, *Aspergillus flavus* and *A. niger* commonly infect the same seed (Wicklow *et al.*, 1987); *A. flavus* produces metabolites toxic to insects, such as the aflatoxins and kojic acid, while *A. niger* produces the toxic alkaloid nigragillin (Wright *et al.*, 1982). It would be interesting to determine the combined action of aflatoxins, nigragillin, and kojic acid on the different insect pests of seeds known to be colonized by these two fungi.

As long as a toxic metabolite critical to defence is included in the total metabolite profile of the fungal species assemblage, then it may not matter which species produces that compound. For example, in temperate climates we find the potent neurotoxic mycotoxin citreoviridin is a product of *Penicillium citreonigrum*, a colonist of rice grain. This fungus grows well at 25°C but shows limited growth or no growth at 37° (Ueno and Ueno, 1972; Pitt, 1980). In subtropical and tropical regions, citreoviridin is a product of another seed colonist, *Eupenicillium ochrosalmoneum* (Wicklow *et al.*, 1982; Horn and Wicklow, 1986). *E. ochrosalmoneum* grows as rapidly at 37°C as it does at 25°C, but less well below 20°C. Here a single metabolite might defend different species of fungi occupying the same niche in different latitudes.

COMPENSATORY RESPONSE TO PREDATION

Compensatory growth or reproductive responses of fungi to moderate grazing may enhance an organism's competitive fitness. Fragmentary evidence suggests that low arthropod grazing intensities can optimize growth and promote sporulation in fungi, thereby increasing an organism's colonizing ability. Findlay (1985) made the exciting discovery that the beneficial effects of mycorrhizal infection to the plant were increased (e.g., total shoot phosphorus; root length, shoot weight) with the addition of fungus-feeding Collembola. He suggested that low grazing intensities may optimize growth of the external mycelium, while at higher Collembolan densities, the effects of fungal feeding outweigh this or any other beneficial effect. Ecologists have come to recognize that plants do not respond passively in the short-term to tissue reduction by herbivory. McNaughton's (1979) review indicates that compensatory growth upon tissue damage by herbivory is a major component of plant adaptation to herbivores. The balance between plant vegetative and reproductive tissues is influenced by herbivory, and fruit and seed yield compensation following herbivory has been reported for many plants. Dyer (1975) has shown that the feeding of red-winged blackbirds on the tips of corn ears during certain developmental stages may stimulate extra growth and seed filling. Dyer theorized that, because potential damage is a function of feeding intensity, there is a point at which the two actions balance out; beyond this, a real crop loss occurs. Is arthropod grazing of fungal mycelium also an optimization process: Brasier (1978) suggests that sporulation of some conidial fungi in response to mechanical injury of the mycelium could also be interpreted as an adaptive response to predation by mites or other invertebrates. Hawker (1966) observed that, while there are several reports of the increased production of conidia in *Fusarium*, *Alternaria*, and other genera resulting from injury to the mycelium, the mechanism by which this stimulatory effect is brought about is unknown.

INTRAFUNGAL DISTRIBUTION OF CHEMICAL DEFENCES

During its life cycle each organism has a finite amount of resources available to it in the form of energy and nutrients. The way in which it partitions these resources between activities (e.g., growth, defence, reproduction) will determine the likelihood of its passing genes of succeeding generations. The principle of allocation (Cody, 1966), is that natural selection results in organisms optimizing the partitioning of their resources to maximize fitness. A plant exposed to a high risk of predation will need to allocate more resources to defence mechanisms such as spines or distasteful chemicals than will a plant that is not so threatened. Such allocation can only be made at the expense of resources devoted to other activities. The actual distribution of resources between activities is presumed to be the optimum compromise brought about by selection. Ecologists have theorized that different types of secondary compounds are differentially suited to the defence of different parts of vascular plants (McKay, 1979; Rhoades, 1979). They are finding that the chemical defence systems of plants are distributed in direct proportion to the risk of the particular tissue and the value of that tissue in terms of fitness loss to the organisms as the result of an attack on that tissue (Rhoades, 1979).

Has predation as a selective force shaped the chemical defence systems of fungi? I addressed this question in connection with the intrafungal distribution of mycotoxins in *A. flavus* (Wicklow and Cole, 1982). Sclerotium-producing fungi such as *A. flavus* are ideally suited for testing hypotheses about the evolution and distribution of fungal chemical defences. A large majority of the fungi that produce sclerotia also produce spores that function in dispersal and, as in the case of plant and insect pathogens, represent the infective inoculum. Because sclerotia are commonly recognized for their nutrient storage, dormancy, and survival properties, efforts to assess their adaptive significance have centred on their resistance to stress in the physical environment, such as desiccation (Willetts, 1971; Coley-Smith and Cooke, 1971), or microbial attack (Lumsden, 1981; Bullock *et al.*, 1986). The survival of sclerotia should also be examined in connection with the means by which these structures deter potential mycophagists (e.g., arthropods, rodents) while lying dormant (Wicklow and Cole, 1982).

Recognizing the potential importance of the fungal sclerotium as a sur-

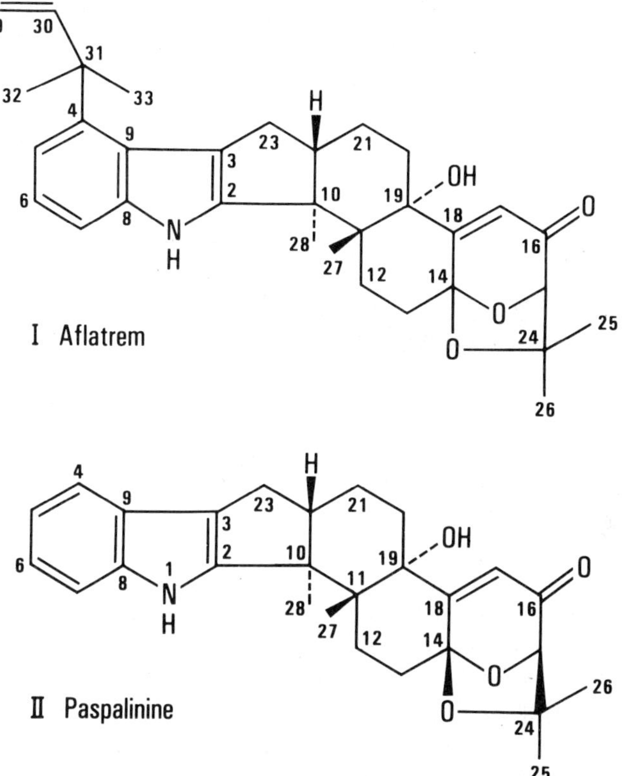

Figure 1. Chemical structures of related tremorgens (I) aflatrem, a metabolite of *Aspergillus flavus*, and (II) paspalinine, a metabolite of *Claviceps paspali*.

vival structure in the life cycle of *A. flavus*, Wicklow and Cole (1982) and Wicklow and Shotwell (1983) predicted that there should be a greater fungal investment (metabolic cost) in sclerotial chemical defences. Aflatoxins B1 and B2 and cyclopiazonic acid were isolated from the sclerotia, conidia, mycelium, and culture medium, whereas aflatoxins G1, G2, aflatrem, and dihydroxyaflavinine were found only in sclerotia harvested from these same cultures. It is significant that aflatrem, from the sclerotium of *A. flavus*, is chemically related to paspalinine (Figure 1), an indole alkaloid metabolite found in the sclerotium of *Claviceps paspali* (Cole, 1981). Wicklow and Cole (1982) theorized that a convergence in metabolic pathways leading to the production of defensive (toxic) secondary compounds would offer different kinds of sclerotia protection from many of the same potential enemies.

Purified sclerotial metabolites from a non-aflatoxigenic strain of *A. flavus* (NRRL 6541) showed substantial anti-feedant activity against *Carpophilus*

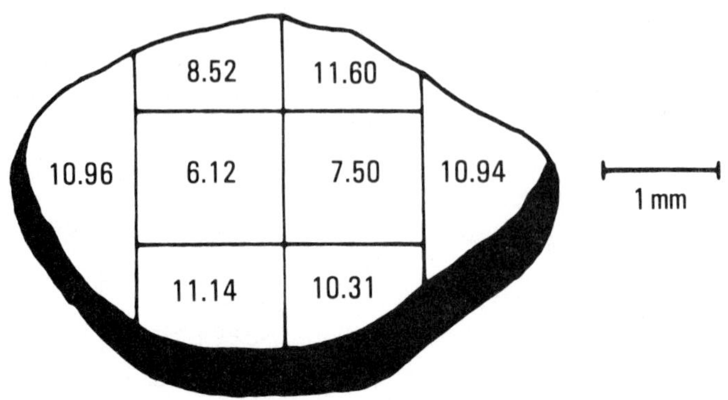

Figure 2. Distribution of ergotamine in the sclerotium of *Claviceps purpurea* (from Arens and Zenk, 1980).

hemipterus (Nitidulidae) (Wicklow *et al*., 1988). This detritivorous insect naturally encounters *A. flavus* sclerotia in both standing and downed maize ears. The most abundant active component was dihydroxyaflavinine, but one novel analogue of dihydroxyaflavinine, and a third unrelated metabolite, were also found to deter *Carpophilus* feeding when incorporated into a pinto bean diet at 100 p.p.m. dry weight. Naturally occurring levels of each of these metabolites were found to be higher than 100 p.p.m. in *A. flavus* sclerotia. Aflatrem, a tremorgen isolated with the hexane extract of *A. flavus* sclerotia, did not elicit an anti-feedant response in *Carpophilus*. It is interesting to speculate whether the evolution of the biosynthetic pathways of these related indole metabolites has been independently guided by both vertebrates and arthropods.

The ergot alkaloids and related indole metabolites found concentrated in the sclerotia of *Claviceps* species (Floss, 1976) probably function in protecting the sclerotium from predation, an ecological role attributed to alkaloids in vascular plant tissues (Robinson, 1979; Clay, this volume, Chapter 4). For example, ergot of *Imperta cylindrica* var. *koenigii*, caused the death of mice in 5 days when they were fed a diet containing 1% (=0.04 g daily) ergot (Tanda *et al*., 1968). Clavine alkaloids produced by *C. fusiformis* infecting millet seed caused loss of lactation in mice and loss of litters in mice when added to the diet (Mantle, 1968). Ergotamine is found in higher concentration in tissue sections nearer to the outer rind of *C. purpurea* sclerotia (Figure 2) than in tissue sections removed from central regions (Arens and Zenk, 1980). A mycophagist would thus encounter the greatest amount of toxin with its first bite, and this might discourage a second bite. That *C. yanagawaensis* sclerotia not found to contain ergot alkaloids were highly poisonous to mice (Tanda, 1981) suggests that sclerotial metabolites with significant biological activity have yet to be characterized. The sclerotia of *Sclerotinia sclerotiorum*, when comprising 5% of the diet of rats, caused a reduced intake of the unpalatable diet, depressed weight gain, and poor body condition (Morrall *et al.*, 1978). Fungal toxins are implicated in each of these examples.

Because fungal sclerotia do not normally form in liquid shake culture, many novel sclerotial metabolites of common moulds such as *Aspergillus* and *Penicillium* may not previously have been detected and characterized. The sclerotia and ascomata produced by many different fungi represent an untapped source of potentially novel secondary metabolites with insecticidal/anti-feedant activity against insect pests.

Garrett (1970) recognized a fundamental distinction between sclerotia produced by airborne fungal pathogens and those produced by root-infecting fungi. The sclerotia of root-infecting fungi are usually small, subspherical, and regular in shape, and function individually and directly as single infective propagules. Infection is by hyphae or hyphal strands (e.g., *Sclerotium rolfsii*, *S. cepivorum*, *Phymatotrichum omnivorum*, and *Verticillium dahliae*). However, the sclerotia of airborne pathogens are variable in size and shape, and upon germination, they produce large numbers of infective propagules in or on fruitbodies that develop from the sclerotia (e.g., *Sclerotinia*, *Typhula*, *Claviceps*). Selective pressure due to predation may effect sclerotium size indirectly, as Janzen (1969) has suggested for seeds. A greater subdivision of the reproductive effort in vulnerable (non-toxic) species, resulting in large numbers of tiny sclerotia as opposed to a few large sclerotia, would offer a greater probability of escape by dispersion. Baker and Cook (1974) recognized that the capacity of a sclerotium to produce secondary inoculum is related to the population density of the fungus necessary to produce severe disease. For example, Holley and Nelson (1986) report that the inoculum densities of *S. sclerotiorum* sclerotia (<1 sclerotium per $800\,cm^3$ of soil) are the lowest reported for a soil-borne pathogenic fungus in terms of propagules per unit of soil and a corresponding disease incidence. Therefore, with the survival of the pathogen linked to the survival of a relatively few sclerotia, one might predict that larger sclerotia should be better chemically defended (i.e., with a number of biosynthetic pathways giving unique classes of compounds) than the tiny sclerotia of root-infecting pathogens. Alternatively, the existence of a size spectrum of sclerotium predators could mean that sclerotium size is not in itself an effective method of avoiding predation and that therefore small sclerotia are also chemically defended (qualitatively and quantitatively) to the same degree as larger sclerotia. Interdisciplinary research is required to test this central hypothesis and answer certain related questions.

1. Is the distribution or allocation of fungal chemical defences a general phenomenon shared by different sclerotium-forming plant pathogenic fungi?
2. Are large sclerotia better protected (in terms of qualitative and quantitative chemical defenses) than smaller sclerotia?
3. Are sclerotia that germinate to produce an air-borne infective inoculum (i.e., conidia, ascospores, basidiospores) better protected regardless of size, than the sclerotia of root-infecting pathogens only capable of myceliogenic germination?
4. Has selection brought about a convergence in the chemical defence systems of sclerotium-producing fungi that are otherwise taxonomically unrelated? Alternatively, because sclerotium-forming ability appears to have evolved independently among distantly related fungal taxa, individual chemical defence systems may differ considerably.
5. Are alkaloids and free amino acids the most satisfactory sclerotial defensive chemicals because of their general toxicity to animals and their potential utility to the germinating sclerotium as nitrogen sources?
6. What is the fate of sclerotial toxins during sclerotium germination? Are some toxins degraded and used in the manufacture of amino acids for growth, or are the toxins merely redistributed to be used in defence of those fungal tissues produced by the germinating sclerotium?

Fungal sclerotia are not immune to arthropod predation. Anas and Reeleder (1987) reported that larvae of the dark-winged fungus gnat (*Bradysia* sp., Sciaridae: Diptera) tunnelled through and fed upon sclerotia of *Sclerotinia sclerotiorum* added to samples of muck soils (organic humic mesisols). The damaged sclerotia had poor levels of myceliogenic germination (0–30%) in contrast to undamaged sclerotia (95% germination). Anas and Reeleder (1987) observed that larvae of *Bradysia* prefer to remain in the top few centimetres of soil and feed on sclerotia in that region. The authors suggest that *Bradysia* sp., and the mycoparasites that invade larval-damaged sclerotia, may play an important role in the rate of natural destruction and lysis of sclerotia. R. D. Reeleder (personal communication) found that *Bradysia* larvae prefer the sclerotia of *S. sclerotiorum* to those of *S. minor* or *Sclerotium cepivorum*. Larvae usually fail to produce "vigorous" adults when provided sclerotia of the latter two fungi as food. This suggests that selective grazing by *Bradysia* could influence the relative abundance of different sclerotium-forming plant pathogenic fungi and thus their inoculum potential. Soil-inhabiting Collembola have been shown to consume the young hyphae of germinating sclerotia of soil-borne plant pathogenic fungi such as *Sclerotium rolfsii*, *Macrophomina phaseolina*, and *Verticillium dahliae*, but apparently cannot readily "destroy" sclerotia (Curl *et al.*, 1985). If we gain an understanding of the sclerotial chemical defence systems of important plant pathogenic fungi, we might design novel biological control strategies based on soil arthropods capable of overcoming (by detoxification) such defences.

Arthropods may also find ascomata less attractive than other propagules produced by the same fungus. Brasier (1978) reported that *Tyrophagus putrescentiae*, a mite isolated from diseased elm bark, ate the mycelium of *Ophiostoma ulmi*, leaving the dark pigmented ascomatal initials and ascomata untouched and intact on twig surfaces. He suggested that the tougher melanized walls of these structures are either physically resistant to the shearing mouth parts of the mites, or possibly chemically repulsive to them. He proposed that the pigmented outer rind of fungal sclerotia and fungal rhizomorphs, often considered a defence against antagonistic microorganisms, are also probably effective in reducing predation.

Chaetomium species may have evolved ornamented ascomatal hairs as a mechanical deterrent and anthraquinones as chemical deterrents to reduce predation by arthropod detritivores (Wicklow, 1979). Larvae of the sciarid fly *Lycoriella mali* avoided ascomata of *Chaetomium bostrychodes* on rabbit faeces, but actively consumed the sporocarps of other coprophilous fungi. In ascomycetes with violent spore discharge mechanisms, the fate of the ascomata following discharge is unimportant. However, *Chaetomium* incorporates a mechanism of spore dispersal that is dependent on the survival of the perithecium that supports an elaborate tuft of terminal hairs upon which the ascospores collect as they are forced from individual perithecia. The presence of larvae of *L. mali* in rabbit faeces decreased the number of sporulating species of coprophilous fungi, and increased the relative abundance of *C. bostrychodes* (Wicklow and Yocom, 1982). Crowson (1984) notes that *Chaetomium* species have no known beetle associates. I suspect that this is related, in part, to the toxicity of *Chaetomium* metabolites, including chaetomin, chaetocins, chaetoglobosins, sterigmatocystin, cochliodinol, and chaetochromin (Udagawa *et al.*, 1979). Pigments of *Chaetomium globosum* acted as chemical sterilants in *Tribolium confusum* (Rao and de las Casas, 1974). Corn invaded by *C. globosum* was lethal to rats (Christensen *et al.*, 1966), and it was established that the toxic agent was chaetomin (Brewer *et al.*, 1972). The metabolites of *Chaetomium* are likely to be targeted against insect predators and should be examined from this perspective. Likewise, in experiments to test the suitability of *Aspergillus* species as food for the grain beetle *Ahasverus advena*, Hill (1978) observed that beetle larvae consumed conidial heads of *Eurotium* but did not eat the ascomata. *Eurotium* ascomata are covered with a felt of sterile aerial hypae encrusted with yellow, orange, or red granules which, if toxic or unpalatable, could make such fungal structures less attractive to arthropods.

In the three-dimensional milieu of soil and litter, sparsely branched mycelial strands and rhizomorphs represent multihyphal structures, which are capable of extending in one direction, enabling a fungus to invade discontinuously distributed nutrient sources (Smith and Berry, 1974; Cooke and Rayner, 1984). Hyphae on the outside of a mycelial cord are frequently heavily encrusted with calcium oxalate crystals (Whitney and Arnott, 1987), which were thought to have an excretory or antibiotic function (Sollins *et al.*, 1981; Cooke and Rayner, 1984). However, a more logical explanation by Thompson (1984) is that crystalline deposits of oxalic acid may protect the hyphal aggregate from litter-inhabiting, fungus-feeding arthropods. The same adaptive characteristics are recognized in species of *Pencillium* that exhibit the funiculose growth habit (Wicklow, 1986).

Because the production of secondary metabolites often coincides with morphological differentiation, such as the formation of spores, the biosynthesis of these metabolites is viewed by some as an aspect of differential gene expression and so of differentiation (Martin and Demain, 1978). Production of secondary metabolites with the formation of aerial mycelium or conidiation has led some students of secondary metabolism to propose that the search for a function for these metabolites should be in the context of the physiology and development of the aerial mycelium or penicillus (Luckner *et al.*, 1977; Campbell, 1983). However, extensive studies on the relationship between conidium formation and secondary metabolism in *Penicillium griseofulvum* (including *P. urticae* and *P. patulum*) indicates that the mycotoxin patulin and related secondary metabolites are not a prerequisite to conidiogenesis (Sekiguchi and Gaucher, 1977). In *P. brevicompactum*, which is common in cereals, the biosynthesis of mycophenolic acid and brevianamide A occurs as aerial hyphae are developing after a vegetative mycelium has formed on solid media (Bartman *et al.*, 1981; Bird *et al.*, 1981; Bird and Campbell, 1982); when grown between dialysis membranes on Czapek-Dox agar, no aerial mycelium forms and no mycophenolic acid or brevianamide A occur (Figure 3), whereas, on a dialysis membrane over the same medium the emergence of aerial mycelium or conidial apparati correlates with the production of these two compounds. Most mycophenolic acid was excreted into the medium (Bird and Campbell, 1982), while brevianamide A appears to be a product of the conidial apparatus (Bird *et al.*, 1981; Bird and Campbell, 1982). The distribution of these secondary metabolites suggests a role in defending both *P. brevicompactum* and the substratum it colonizes from predators. Mycophenolic acid is toxic to vertebrates and may discourage seed consumption, but brevianamide A was not toxic when doses up to 40 mg were given to mice (Wilson *et al.*, 1973). Given its localized distribution in the penicillus, brevianamide A may deter fungivorous arthropods from consuming the penicillus (Wicklow, 1986). Support for this hypothesis is provided by

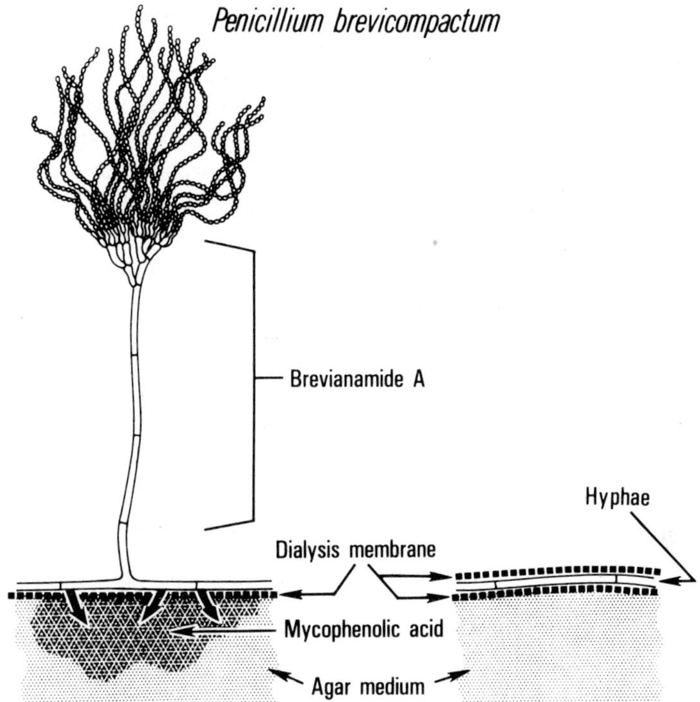

Figure 3. Production and distribution of mycophenolic acid and brevianamide A by sporulating (over dialysis membrane) and non-sporulating (between dialysis membranes) cultures of *Penicillium brevicompactum*, × 100.

the first valid toxicity data attributed to brevianamide A (Paterson *et al.*, 1987); high levels of feeding inhibition against second-instar larvae of *Drosophila melanogaster*, and substantial mortality of fourth-instar larvae of *Spodoptera littoralis*, occurred when brevianamide A was incorporated into the diet at 10 p.p.m. Crude extracts from sporulating cultures of *P. brevicompactum* producing brevianamide A were also toxic. The toxicity of forage colonized by *Pithomyces chartarum* is apparently correlated with the amount of sporulation (conidia) of the fungus on that forage. The conidia contain spordesmins, the toxins responsible for the mycotoxicosis (Taylor, 1971; Atherton *et al.*, 1974). The toxicity of the forage could therefore be related to the ingestion of these conidia; if these conidia were removed, the forage would not then remain toxic. Efforts by chemists to extract and purify individual metabolites following fermentation can also provide information about whether a metabolite is retained by the fungal hyphae or passed into the surrounding medium (Purchase, 1974; Filtenborg *et al.*, 1983). Such data could prove useful to chemical ecologists in developing hypotheses as to the role of individual toxic metabolites in the chemical defence of the producing organisms.

The potential morphological defences associated with the conidial apparati of conidial fungi should not be overlooked (Wicklow, 1986). Fasciculate pencilllia that have strongly roughened conidiophores but smooth conidia (e.g., *Penicillium viridicatum, P. aurantiogriseum*) are among the most commonly encountered toxigenic moulds in foods or feeds stored below 25°C. While a coarsely roughened conidiophore could discourage microarthropod fungivores, the smooth conidia should not. Fungus-feeding arthropods are important in the dispersal of storage moulds within grain bulks (Sinha *et al.*, 1969). The more successful of these are likely to have evolved different morphological and chemical characteristics to facilitate their dispersal by arthropods. Smooth conidia are also produced by *P. chrysogenum* var. *dipodomyis*, distinguished in part by its roughened conidiophores, and known only from the seed caches of the Sonoran Desert rodent *Dipodomys spectabilis* (Frisvad *et al.*, 1987). Seastedt *et al.* (1985) reported large numbers of fungivorous microarthropods in these same rodent caches; roughened stipes could represent an important adaptation in this habitat.

FUNGAL TOXINS IN BIOCONTROL

The design and interpretation of experiments using fungal seed dressings in biocontrol (Harman, 1983; Kommedahl and Windels, 1978, 1981) should recognize the role of soil invertebrates in producing entry wounds for soil-borne seed-infecting plant pathogenic fungi. Feeding by soil invertebrates on seeds or roots can provide access for soil-borne fungal pathogens of crop plants (Agarios, 1980). For example, root rot in barley caused by *Bipolaris sorokiniana* is positively correlated with the numbers of stunt nematodes (*Tylenchorhynchus* spp.) in soil (Kimpinski and Johnston, 1985). There are at the same time numerous reports that specific fungal antagonists, applied as spores in seed dressings, can serve as effective biocontrol agents against these pathogens (Baker and Cook, 1974; Kommedahl and Windels, 1981; Tveit and Moore, 1954; Tveit and Wood, 1955). It is important to determine whether the protection provided by such fungal antagonists results, in part, from interference with invertebrate feeding. With reference to the above example, we might learn if some biocontrol fungi, by virtue of their metabolites, are better able to deter feeding by stunt nematodes than others. I do not consider it an accident that those fungi reported to be the most effective seed dressings (i.e., *Penicillium oxalicum*, *Trichoderma viride*, *Chaetomium globosum*) produce metabolites toxic to animals. The results of Harman and coworkers support this hypothesis. Seed-borne microorganisms on a Butternut squash (*Cucurbita pepo*) stimulate seedcorn maggot egg laying (Eckenrode *et al.*, 1975), but, if the seeds are first dusted with ascospores of *Chaetomium globosum*, this negatively affects oviposition by this insect, reduces larval damage to seeds, and prevents subsequent attack by soil-borne fungi (e.g., *Pythium* species) on germinating seeds (Harman *et al.*, 1978). Fungi applied as seed dressings are able to colonize the seed surface heavily (Windels, 1981). The sporulating fungus is also capable of producing toxic metabolites in this environment (Wright, 1956). The seed corn maggot avoids seeds colonized by *C. globosum*, I presume because of the numerous toxic metabolites this fungus produces (Udagawa *et al.*, 1979). It is easy to envisage how there might be substantial variation in the performance of biocontrol treatments with antagonistic microorganisms (Kommedahl *et al.*, 1981), if the conditions for toxin production are not met in the spermatosphere of certain seed types, or if certain arthropods are not adversely affected by the toxins of the biocontrol fungus.

Do seeds in the wild protect themselves from predators by harbouring fungi that manufacture toxic metabolites? Harborne (1977) suggested that some higher plants may protect themselves by acquiring fungal toxins in their defence. Clay (1984; and this volume, Chapter 4) has shown that individuals of the grass *Danthonia spicata* infected by *Atkinsonella hypoxylon* had greater fitness than uninfected individuals in field trials. This fungus produces alkaloids similar to the ergot alkaloids of *Claviceps* species and Clay argues that since the fungal endophyte increases the fitness of its host then mutualism is in effect. Seeds often support diverse fungal colonists prior to dispersal (Neergaard, 1977; Noble *et al.*, 1958; Agarwal and Sinclair, 1987). These fungi may be internally seed-borne and asymptomatic, or on the seed surface. Some are toxin producers (e.g., *Fusarium moniliforme*, *Penicillium oxalicum*, *Chaetomium globosum*), while others may trap nematodes (e.g., *Arthrobotrys* spp.) or function as mycoparasites (e.g., *Penicillium funiculosum*). In addition, *P. funiculosum* promotes the germination of *Oryzopsis miliacea* (Probert, 1981). The various abilities to deter potential seed pathogens or predators suggest that selection may be operating in the formation of such species assemblages. Indeed, Baker and Cook (1974) have argued that the most effective microbial inoculum for the biocontrol of soil-borne plant pathogens would be a combination of organisms with differing abilities to function as antagonists under different environmental conditions.

SUMMARY

In this chapter, I have attempted to show how risk value and cost are involved in shaping the chemical defence systems of fungi. It is important to recognize that fungi may produce specific metabolites in association with the formation of different propagule types (e.g., rhizomorphs, conidia, sclerotia). There is a wealth of data on the secondary metabolite profiles of fungal taxa that are important to agriculture. Multidisciplinary teams comprising ecologically orientated mycologists, entomologists, toxicologists, and natural-product chemists have begun to characterize these fungal chemical defence systems. It is important that this current research effort focuses on the interactions of toxigenic fungi with animals living in the same habitat, and that investigators attempt to examine their results within the broader ecological context of plant–herbivore theory.

ACKNOWLEDGEMENTS

Professor Richard P. Seifert (1947–81), my late friend and colleague, is remembered for inspiring an evolutionary ecologist's view of the mycological world.

REFERENCES

Agarwal, V. K. and Sinclair, J. B. (1987). "Principles of Seed Pathology". CRC Press, Boca Raton, Fl.

Agrios, G. N. (1980). Insect involvement in the transmission of fungal pathogens. In "Vectors of Plant Pathogens" (K. F. Harris and K. Maramorosch, eds), pp. 293–324. Academic Press, New York.

Anas, O. and Reeleder, R. D. (1987). Recovery of fungi and arthropods from sclerotia of *Sclerotinia sclerotiorum* in Quebec muck soils. *Phytopathology* **77,** 327–331.

Anderson, J. M. and Ineson, P. (1984). Interactions between microorganisms and soil invertebrates in nutrient flux pathways of forest ecosystems. In "Invertebrate–Microbial Interactions" (J. M. Anderson, A. D. M. Rayner, and D. W. H. Walton, eds), pp. 59–88. Cambridge University Press, Cambridge.

Arai, T., Ito, T., and Koyama, Y. (1967). Antimicrobial activity of aflatoxins. *J. Bacteriol.* **93,** 59–64.

Arens, H. and Zenk, M. H. (1980). Radioimmuntests fur die bestimmung der peptidealkaloide ergotamin and ergocristin in *Claviceps purpurea. Pl. med.* **38,** 214–226.

Arsuffi, T. L. and Suberkropp, K. (1985). Selective feeding by stream caddis fly (Trichoptera) detritivores on leaves with fungal-colonized patches. *Oikos* **45,** 50–58.

Atherton, L. G., Brewer, D., and Taylor, A. (1974). *Pithomyces chartarum*: a fungal parameter in the etiology of some diseases of domestic animals. In "Mycotoxins" (I. F. H. Purchase, ed.), pp. 29–68. Elsevier, Amsterdam.

Atkinson, W. D. (1981). An ecological interaction between citrus fruit, *Penicillium* moulds and *Drosophila immigrans* Sturtevant (Diptera: Drosophilidae). *Ecol. Entom.* **6,** 339–344.

Baker, K. F. and Cook, R. J. (1974). "Biological Control of Plant Pathogens". Freeman, San Francisco.

Bartman, C. D., Doerfler, D. I., Bird, B. A., Remaley, A. T., Peace, J. N., and Campbell, I. M.

(1981). Mycophenolic acid production by *Penicillium brevicompactum* on solid media. *Appl. env. Microbiol.* **41,** 729–736.

Bennett, J. W. (1981). Genetic perspective on polyketides, productivity, parasexuality, protoplasts, and plastids. In "Advances in Biotechnology", Vol. 3. "Fermentation Products" (C. Vezina and K. Singh, eds), pp. 409–415. Pergamon Press, Toronto.

Bennett, J. W. (1983). Differentiation and secondary metabolism in mycelial fungi. In "Differentiation and Secondary Metabolism in Fungi" (J. W. Bennett and A. Ciegler, eds), pp. 1–32. Marcel Dekker, New York and Basel.

Berenbaum, M. and Neal, J. J. (1985). Synergism between myristicin and xanthotoxin, a naturally occurring plant toxicant. *J. Chem. Ecol.* **11,** 1349–1358.

Bernays, E. A. and Woodhead, S. (1982). Plant phenols utilized as nutrients by a phytophagous insect. *Science, N.Y.* **216,** 201–202.

Bird, B. A. and Campbell, I. M. (1982). Disposition of mycophenolic acid brevianamide A, asperphenamate, and ergosterol in solid cultures of *Penicillium brevicompactum. Appl. env. Microbiol.* **43,** 345–348.

Bird, B. A., Remaley, A. T., and Campbell, I. M. (1981). Brevianamides A and B are formed only after conidiation has begun in solid cultures of *Penicillium breviocompactum. Appl. env. Microbiol.* **42,** 521–525.

Brasier, C. M. (1978). Mites and reproduction in *Ceratocystis ulmi* and other fungi. *Trans. Br. mycol. Soc.* **70,** 81–89.

Brewer, D., Duncan, J. M., Jerram, W. A., Leach, C. K., Safe, S., Taylor, A., Vining, L. C., Archibald, R. Mc G., Stevenson, R. G., Mirocha, C. J., and Christensen, C. M. (1972). Ovine ill-thrift in Nova Scotia. 5. The production and toxicology of chaetomin, a metabolite of *Chaetomium* spp. *Can. J. Microbiol.* **18,** 1129–1137.

Bruns, T. (1984). Insect mycophagy in the Boletales: fungivore diversity and the mushroom habitat. In "Fungus–Insect Relationships" (Q. Wheeler and M. Blackwell, eds), pp. 91–129. Columbia University Press, New York.

Bullock, S., Adams, P. B., Willetts, H. J., and Ayers, W. A. (1986). Production of haustoria by *Sporidesmium sclerotivorum* in sclerotia of *Sclerotinia minor. Phytopathology* **76,** 101–103.

Bu'Lock, J. D. (1980). Mycotoxins as secondary metabolites. In "The Biosynthesis of Mycotoxins: A study in Secondary Metabolism" (P. S. Steyn, ed.), pp. 1–16. Academic Press, New York.

Burmeister, H. R. and Hesseltine, C. W. (1966). Survey of the sensitivity of microorganisms to aflatoxin. *Appl. Microbiol.* **14,** 403–404.

Calam, C. T. (1979). Secondary metabolism as an expression of microbial growth and development. *Folia microbiol.* **24,** 276–285.

Campbell, I. M. (1983). Correlation of secondary metabolism and differentiation. In "Secondary Metabolism and Differentiation in Fungi" (J. W. Bennett and A. Ciegler, eds), pp. 55–72. Marcel Dekker, New York and Basel.

Charnley, A. K. (1984). Physiological aspects of destructive pathogenesis in insects by fungi: a speculative review. In "Invertebrate–Microbial Interactions" (J. M. Anderson, A. D. M. Rayner and D. W. H. Walton, eds), pp. 229–270. Cambridge University Press, London.

Christensen, C. M., Nelson, G. H., Mirocha, C. J., Bates, F., and Dorworth, C. E. (1986). Toxicity to rats of corn invaded by *Chaetomium globosum. Appl. Microbiol.* **14,** 774–777.

Ciegler, A. (1983). Evolution, ecology, and mycotoxins: some musings. In "Secondary Metabolism and Differentiation in Fungi" (J. W. Bennett and A. Ciegler, eds), pp. 429–439. Marcel Dekker, New York and Basel.

Clay K. (1984). The effect of the fungus *Atkinsonella hypoxylon* (Clavicipitaceae) on the reproductive system and demography of the grass *Danthonia spicata. New Phytol.* **98,** 165–175.

Cody, M. L. (1966). A general theory of clutch size. *Evolution* **20,** 174–184.

Cole, R. J. (1981). Fungal tremorgens. *J. Fd Prot.* **44,** 715–722.

Coley-Smith, J. R. and Cooke, R. C. (1971). Survival and germination of fungal sclerotia. *A. Rev. Phytopath.* **9,** 65–92.

Cooke, R. C. and Rayner, A. D. M. (1984). "Ecology of Saprotrophic Fungi". Longman, New York.

Crowson, R. A. (1984). The association of Coleoptera with Ascomycetes. In "Fungus–Insect Relationships" (Q. Wheeler and M. Blackwell, eds), pp. 256–285. Columbia University Press, New York.

Curl, E. A., Gudauskas, R. T., and Peterson, C. M. (1985). Effects of soil insects on populations and germination of fungal propagules. In "Ecology and Management of Soil-borne Plant Pathogens" (C. A. Parker, A. D. Rovira, K. J. Moore, P. T. W. Wong and J. F. Kollmorgen, eds), pp. 20–23. American Phytopathological Society, St Paul, Minn.

Dingley, J. M., Done, J., Taylor, A., and Russell, D. W. (1962). The production of sporidesmin and sporidesmolides by wild isolates of *Pithomyces chartarum* in surface and in submerged culture. *J. gen. Microbiol.* **29,** 127–135.

Dowd, P. F. (1988a). Synergism of aflatoxin B1 toxicity with the co-occurring fungal metabolite kojic acid to two caterpillars. *Entomol. Exper. appl.*, in press.

Dowd, P. F. (1988b). Fusaric acid: a secondary fungal metabolite that synergizes the toxicity of coocurring host allelochemicals to the corn earworm, *Heliothis zea* (Lepidoptera). *J. Chem. Ecol.*, in press.

Dowd, P. F., Smith, C. M. and Sparks, T. C. (1983). Detoxification of plant toxins by insects. *Insect Biochem.* **13,** 453–468.

Dyer, M. I. (1975). The effects of red-winged blackbirds (*Agelaius phoeniceus* L.) on biomass production of corn grains (*Zea mays* L.). *J. appl. Ecol.* **12,** 719–726.

Eckenrode, C. J., Harman, G. E. and Webb, D. R. (1975). Seed-borne microorganisms stimulate seedcorn maggot egg laying. *Nature, Lond.* **256,** 487–488.

Filtenborg, O., Frisvad, J. C., and Svendsen, J. A. (1983). Simple screening methods for molds producing intracellular mycotoxins in pure cultures. *Appl. env. Microbiol.* **45,** 581–585.

Findlay, R. D. (1985). Interactions between soil micro-arthropods and endomycorrhizal associations of higher plants. In "Ecological Interactions in Soil" (A. H. Fitter, ed.), pp. 319–332. Blackwell Scientific, Oxford.

Floss, H. G. (1976). Biosynthesis of ergot alkaloids and related compounds. *Tetrahedron* **32,** 873–912.

Forgacs, J. (1972). Stachybotryotoxicosis. In "Microbial Toxins". Vol. 8. "Fungal Toxins" (S. Kadis, A. Ciegler and S. J. Ajl, eds), pp. 95–128. Academic Press, New York.

Frisvad, J. C. (1986). Taxonomic approaches to mycotoxin identification (Taxonomic indication of mycotoxin content in foods). In "Modern Methods in the Analysis and Structural Elucidation of Mycotoxins" (R. J. Cole, ed.), pp. 415–457. Academic Press, New York.

Frisvad, J. C., Filtenborg, O., and Wicklow, D. T. (1987). Terverticillate penicillia isolated from underground seed caches and cheek pouches of banner-tailed kangaroo rats (*Dipodomys spectabilis*). *Can. J. Bot.* **65,** 765–773.

Garrett, S. D. (1970). "Pathogenic Root-infecting Fungi". Cambridge University Press, Cambridge.

Harborne, J. B. (1977). "Introduction to Ecological Biochemistry". Academic Press, New York.

Harman, G. E. (1983). Mechanisms of seed infection and pathogenesis. *Phytopathology* **73,** 326–329.

Harman, G. E., Eckenrode, C. J., and Webb, D. R. (1978). Alteration of spermatosphere ecosystems affecting oviposition by the bean seed fly and attack by soilborne fungi on germinating seeds. *Ann. appl. Biol.* **90,** 1–6.

Hawker, L. E. (1966). Environmental influences on reproduction. In "The Fungi: An Advanced

Treatise" (G. C. Ainsworth and A. S. Sussman, eds), vol. 2, pp. 435–469. Academic Press, New York.

Hawksworth, D. L. and Hill, D. J. (1984). "The Lichen-Forming Fungi". Blackie, Glasgow and London.

Hill, S. T. (1978). Development of *Ahasverus advena* (Coleoptera: Silvanidae) on seven species of *Aspergillus* and on food molded by two of these. *J. stored Prod. Res.* **14,** 227–231.

Holley, R. C. and Nelson, B. D. (1986). Effect of plant population and inoculum density on incidence of *Sclerotinia* wilt of sunflower. *Phytopathology* **76,** 71–74.

Horn, B. W. and Wicklow, D. T. (1986). Ripening of *Eupenicillium ochrosalmoneum* ascostromata on soil. *Mycologia* **78,** 248–252.

Jaenike, J. (1985). Parasite pressure and the evolution of amanitin tolerance in *Drosophila. Evolution* **39,** 1295–1301.

Janzen, D. H. (1969). Seed-eaters versus seed size, number, toxicity and dispersal. *Evolution* **23,** 1–27.

Janzen, D. H. (1977). Why fruits rot, seeds mold and meat spoils. *Am. Nat.* **111,** 691–713.

Janzen, D. H. (1978). The ecology and evolutionary biology of seed chemistry as relates to seed predation. In "Biochemical Aspects of Plant and Animal Coevolution" (J. B. Harborne, ed.), pp. 163–206. Academic Press, New York.

Jarvis, J. L., Guthrie, W. D., and Lillehoj, E. B. (1984). Aflatoxin and selected biosynthetic precursors: effects on the European corn borer in the laboratory. *J. agric. Entomol.* **1,** 354–359.

Jones, J. G. (1981). Models of speciation—the evidence from *Drosophila. Nature, Lond.* **289,** 743–744.

Karban, R., Adamchak, R., and Schnathorst, W. C. (1987). Induced resistance and interspecific competition between spider mites and a vascular wilt fungus. *Science, N.Y.* **235,** 678–680.

Kobel, H. and Sanglier, J. J. (1978). Formation of ergotoxine alkaloids by fermentation and attempts to control their biosynthesis. In "Antibiotics and Other Secondary Metabolites: Biosynthesis and Production" (R. Hutter, T. Leisinger, J. Neusch, and W. Wehrli, eds). pp. 233–242. Academic Press, London.

Kimpinski, J. and Johnston, H. W. (1985). Incidence of root rot and nematodes in barley fields in Prince Edward Island (Canada). *Can. Pl. Dis. Serv.* **65,** 15–16.

Kingsley, P., Scriber, J. M., Grau, C. R., and Delwiche, P. A. (1983). Feeding and growth performance of *Spodoptera eridania* (Noctuidae: Lepidoptera) on "vernal" alfalfa, as influenced by *Verticillium* wilt. *Protection Ecology* **5,** 127–134.

Kommedahl, T. and Windels, C. E. (1978). Evaluation of biological seed treatment for controlling root diseases of pea. *Phytopathology* **68,** 1087–1095.

Kommedahl, T. and Windels, C. E. (1981). Introduction of microbial antagonists to specific courts of infection: Seeds, seedling, and wounds. In "Biological Control in Crop Protection" (G. Papavizas, ed.), pp. 227–248. Allanheld Osmun, Totowa, N.J.

Kommedahl, T., Windels, C. E., Sarbini, G., and Wiley, H. B. (1981). Variability in performance of biological and fungicidal seed treatments in corn, peas, and soybeans. *Protection Ecology* **3,** 55–61.

Lawrey, J. D. (1980). Correlations between lichen secondary chemistry and grazing activity by *Pallifera varia. Bryologist* **83,** 328–334.

Levin, D. A. (1978). Alkaloids and geography. *Am. Nat.* **112,** 1133–1134.

Lillehoj, E. B. (1981). Secondary metabolites as chemical signals between species in an ecological niche. In "Advances in Biotechnology", Vol. 3. "Fermentation Products". (C. Venzina and K. Singh, eds), pp. 397–402. Pergamon Press, Toronto.

Loschiavo, S. R. and Sinha, R. N. (1966). Feeding, oviposition, aggregation by the rusty grain beetle *Cryptolestes ferrugineus* (Coleoptera: Cucujidae) on seed-borne fungi. *Ann. ent. Soc. Am.* **59,** 578–585.

Luckner, M., Nover, L., and Bohm, H. (1977). Secondary metabolism and cell differentiation. *Mol. Biol. Biochem. Biophys.* **23,** 3–102.

Lumsden, R. D. (1981). Ecology of mycoparasitism. In "The Fungal Community, Its Organization and Role in the Ecosystem" (D. T. Wicklow and G. C. Carroll, eds), pp. 295–318. Marcel Dekker, New York and Basel.

Manabe, M., and Tsuruta, O. (1978). Geographical distribution of aflatoxin-producing fungi inhabiting in southeast Asia. *Jap. agric. Res. Q.* **12,** 224–227.

Madelin, M. F. (1963). Diseases caused by hyphomycetous fungi. In "Insect Pathology: An Advance Treatise" (E. A. Steinhaus, ed.), vol. 2, pp. 233–271. Academic Press, New York.

Mantle, P. G. (1968). Inhibition of lactation in mice following feeding with ergot sclerotia (*Claviceps fusiformis* Loveless) from the bulrush millet (*Pennisetum typhoides* Staph and Hubbard) and an alkaloid component. *Proc. R. Soc. B* **170,** 423–434.

Martin, J. F. and Demain, A. L. (1978). Fungal development and metabolite formation in filamentous fungi. In "Developmental Mycology" (J. E. Smith and D. R. Berry, eds), pp. 426–451. Wiley, New York.

Martin, M. (1979). Biochemical implications of insect mycophagy. *Biol. Rev.* **54,** 1–21.

McKay, D. (1979). The distribution of secondary compounds within plants. In "Herbivores: Their Interaction with Secondary Plant Metabolites" (G. A. Rosenthal and D. H. Janzen, eds), pp. 55–133. Academic Press, New York.

McNaughton, S. J. (1979). Grazing as an optimization process: Grass-ungulate relationships in the Serengeti. *Am. Nat.* **113,** 691–703.

Morrall, R. A. A., Loew, F. M. and Hayes, M. A. (1978). Subacute toxicological evaluation of sclerotia of *Sclerotinia sclerotiorum* in rats. *Can. J. comp. Med.* **42,** 473–477.

Neergard, P. (1977). "Seed Pathology". 2 vols. Halsted Press, Wiley, New York.

Newell, K. (1984a). Interactions between two decomposer basidiomycetes and a Collembolan under Sitka Spruce: distribution, abundance and selective grazing. *Soil Biol. Bochem.* **16,** 235–239.

Newell, K. (1984b). Interactions between two decomposer basidiomycetes and a Collembolan under Sitka Spruce: grazing and its potential effects on fungal distribution and litter decomposition. *Soil Biol. Biochem.* **16,** 235–239.

Noble, M., De Tempe, J., and Neergaard, P. (1958). "An Annotated List of Seed-borne Diseases". Commonwealth Mycological Institute, Kew.

Parkinson, D., Visser, S., and Whittaker, J. B. (1979). Effects of Collembolan grazing on fungal colonisation of leaf litter. *Soil Biol. Biochem.* **11,** 529–535.

Paterson, R. R. M., Simmonds, M. S. J., and Blaney, W. M. (1987). Mycopesticidal effects of characterized extracts of *Penicillium* isolates and purified secondary metabolites (including mycotoxins) on *Drosophila melanogaster* and *Spodoptora littoralis*. *J. Invertebr. Path.* **50,** 124–133.

Pitt, J. I. (1980) ["1979"]. "The Genus *Penicillium* and its Teleomorphic States *Eupenicillium* and *Talaromyces*". Academic Press, London.

Probert, R. J. (1981). The promotive effects of a mould, *Penicillium funiculosum* Thom on the germination of *Oryzopsis miliacea* (L.). Asch et Schw. *Ann. Bot.* **48,** 85–88.

Purchase, I. H. F. (1974). *Penicillium cyclopium*. In "Mycotoxins" (I. H. F. Purchase, ed.), pp. 149–162. Elsevier, Amsterdam.

Rao, H. R. G. and de las Casas, E. (1974). Effect of metabolites of *Chaetomium* spp. on *Tribolium confusum*. *Proc. N. cent. Brch. Am. Ass. econ. Ent.* **29,** 152–153.

Raper, K. B. and Fennell, D. I. (1965). "The Genus *Aspergillus*". Williams and Wilkins, Baltimore.

Rawlins, J. (1984). Mycophagy in Lepidoptera. In "Fungus–Insect Relationships" (Q. Wheeler and M. Blackwell, eds), pp. 382–423. Columbia University Press, New York.

Rhoades, D. F. (1979). Evolution of plant chemical defense against herbivores. In "Herbivores: Their Interaction with Secondary Plant Metabolites" (G. A. Rosenthal and D. H. Janzen, eds), pp. 3–54. Academic Press, New York.

Rhoades, D. F. (1985). Offensive-defensive interactions between herbivores and plants: their relevance in herbivore population dynamics and ecological theory. *Am. Nat.* **125,** 205–238.

Roberts, D. W. (1966). Toxins from the entomogenous fungus *Metarhizium anisopliae*. II. Symptoms and detection in moribund hosts. *J. Invertebr. Path.* **8,** 222–227.

Robinson, T. (1979). The evolutionary ecology of alkaloids. In "Herbivores: Their Interaction with Secondary Plant Metabolites" (G. A. Rosenthal and D. H. Janzen, eds), pp. 413–448. Academic Press, New York.

Rosenthal, G. A., Janzen, D. H., and Dahlman, D. L. (1977). Degradation and detoxification of canavanine by a specialized seed predator. *Science, N.Y.* **196,** 658–660.

Rothschild, M. (1973). Secondary plant substances and warning colouration in insects. In "Insect Plant Relationships" (H. F. van Emden, ed.) pp. 59–83. Blackwell, London.

Seastedt, T. R., Reichman, O. J. and Todd, T. C. (1985). Microarthropods and nematodes in kangaroo rat burrows. *Southwest. Nat.* **31,** 114–116.

Sekiguchi, J. and Gaucher, G. M. (1977). Conidiogensis and secondary metabolism in *Penicillium urticae*. *Appl. env. Microbiol.* **33,** 147–158.

Shaw, P. J. A. (1985). Grazing preferences of *Onchiurus armatus* (Insecta: Collembola) for mycorrhizal and saprophytic fungi of pine plantations. In "Ecological Interactions in Soil" (A. H. Fitter, D. Atkinson, D. J. Read and M. B. Usher, eds), pp. 333–337. Blackwell, Oxford.

Sinha, R. N. (1964). Ecological relationships of stored product mites and seed-borne fungi. *Acarologia* **6,** 372–389.

Sinha, R. N. (1966). Feeding and reproduction of two species of stored product beetles on selected fungi. *J. Invertebr. Path.* **33,** 115–117.

Sinha, R. N. (1971). Fungus as food for some stored product insects. *J. econ. Entomol.* **64,** 3–6.

Sinha, R. N., Wallace, H. A. H., and Chebib, F. S. (1969). Principal component analysis of interrelations among fungi, mites and insects in grain bulk ecosystems. *Ecology* **50,** 536–547.

Smith, J. E. and Berry, D. R. (1974). "An Introduction to the Biochemistry of Fungal Development". Academic Press, London.

Sollins, P., Cromack, K., Fogel, R., and Li, C. Y. (1981). Role of low-molecular weight organic acids in the inorganic nutrition of fungi and higher plants. In "The Fungal Community, its organization and role in the ecosystem" (D. T. Wicklow and G. C. Carroll, eds), pp. 607–629. Marcel Dekker, New York and Basel.

Steiner, W. F. (1984). A review of the biology of Phalacrid beetles. In "Fungus–Insect Relationships" (Q. Wheeler and M. Blackwell, eds), pp. 424–445. Columbia University Press, New York.

Tamura, S. and Takahashi, N. (1971). Destruxins and piericidins. In "Naturally Occurring Insecticides" (M. Jacobson and D. G. Crosby, eds), pp. 499–539. Marcel Dekker, New York.

Tanda, S. (1981). Mycological studies on the ergot of Japan. XX. *Claviceps yanagawwaensis* Togashi parasitic on *Zoysia japonica* Steud. *J. agric. Sci. Tokyo* **26,** 193–199.

Tanda, S., Tadakuma, Y., and Matsunami, Y. (1968). The fundamental studies on ergotial fungi. VIII. Poisonous experiments of ergot to mouse. *J. agric. Sci. Tokyo* **13,** 55–60.

Taylor, A. (1971). The toxicology of sporidesmins and other epipolythiadioxpiperazines. In "Microbial Toxins", Vol. 8. "Algal and Fungal Toxins" (S. Kadis, A. Ciegler and S. J. Ajl, eds), pp. 337–376. Academic Press, New York.

Thompson, W. (1984). Distribution, development and functioning of mycelial and cord systems of decomposer basidiomycetes of the deciduous woodland floor. In "The Ecology and

Physiology of the Fungal Mycelium" (D. H. Jennings and A. D. M. Rayner, eds), pp. 185–214. Cambridge University Press, London.

Tveit, M. and Moore, M. B. (1954). Isolates of *Chaetomium* that protect oats from *Helminthosporium victoriae*. *Phytopathology* **44,** 686–689.

Tveit, M. and Wood, R. K. S. (1955). The control of *Fusarium* blight in oat seedlings with antagonistic species of *Chaetomium*. *Ann. appl. Biol.* **43,** 538–552.

Udagawa, S., Muroi, T., Kurata, H., Sexita, S., Yoshihira, K., Natori, S., and Umeda, M. (1979). The production of chaetoglobosins, sterigmatocystin, o-methylsterigmatocystin, and chaetocin by *Chaetomium* spp. and related fungi. *Can. J. Microbiol.* **25,** 170–177.

Ueno, Y. and Ueno, I. (1972). Isolation and acute toxicity of citreviridin, a neurotoxic mycotoxin of *Penicillium citreo-viride* Biourge. *Jap. J. exp. Med.* **42,** 91–105.

van Bronswijk, J. E. M. H. and Sinha, R. N. (1971). Interrelations among physical, biological and chemical variates in stored-grain ecosystems: a descriptive and multivariate study. *Annals ent. Soc. Am.* **64,** 789–803.

Whitney, K. D. and Arnott, H. J. (1987). Calcium oxalate crystal morphology and development in *Agaricus bisporus*. *Mycologia* **79,** 180–187.

Whittaker, R. H. and Feeney, P. (1971). Allelochemics: chemical interactions between species. *Science, N.Y.* **171,** 757–770.

Wicklow, D. T. (1979). Hair ornamentation and predator defense in *Chaetomium*. *Trans. Br. mycol. Soc.* **72,** 107–110.

Wicklow, D. T. (1981). Interference competition and the organization of fungal communities. In "The Fungal Community, its organization and role in the ecosystem" (D. T. Wicklow and G. C. Carroll, eds), pp. 351–375. Marcel Dekker, New York and Basel.

Wicklow, D. T. (1984). Adaptation in wild and domesticated yellow-green aspergilli. In "Toxigenic Fungi—Their Toxins ahd Health Hazard" (H. Kurata and Y. Ueno, eds), pp. 78–86. Elsevier, Amsterdam.

Wicklow, D. T. (1986) ["1985"]. Ecological adaptations and classification in *Aspergillus* and *Penicillium*. In "Advances in *Penicillium* and *Aspergillus* Systematics" (R. A. Samson and J. I. Pitt, eds), pp. 255–265. Plenum Publishing, New York.

Wicklow, D. T. and Cole, R. J. (1982). Tremorgenic indole metabolites and aflatoxins in sclerotia of *Aspergillus flavus* Link: an evolutionary perspective. *Can. J. Bot.* **60,** 525–528.

Wicklow, D. T. and Shotwell, O. L. (1983). Intrafungal distribution of aflatoxin among conidia and sclerotia of *Aspergillus flavus* and *Aspergillus parasiticus*. *Can. J. Microbiol.* **29,** 1–5.

Wicklow, D. T. and Yocom, D. H. (1982). The effect of larval grazing by *Lycoriella mali* (Diptera: Sciaridae) on species abundance of coprophilous fungi. *Trans. Br. mycol. Soc.* **78,** 29–32.

Wicklow, D. T., Horn, B. W. and Cole, R. J. (1982). Cleistothecia of *Eupenicillium ochrosalmoneum* form naturally within corn kernels. *Can. J. Bot.* **60,** 1050–1053.

Wicklow, D. T., Horn, B. W., and Shotwell, O. L. (1987). Aflatoxin formation in preharvest maize ears coinoculated with *Aspergillus flavus* and *Aspergillus niger*. *Mycologia* **79,** 679–682.

Wicklow, D. T., Dowd, P. F., Tepaske, M. R., and Gloer, J. B. (1988). Sclerotial metabolites of *Aspergillus flavus* toxic to a detritivorous maize insect (*Carpophilus hemipterus*, Nitidulidae). *Trans. Br. mycol. Soc.*, in press.

Willetts, H. J. (1971). The survival of fungal sclerotia under adverse conditions. *Biol. Rev.* **46,** 387–407.

Wilson, B. J., Yang, D. T. C., and Harris, T. M (1973). Production, isolation, and preliminary toxicity studies of brevianamide A from cultures of *Penicillium viridicatum*. *Appl. Microbiol.* **26,** 633–635.

Windels, C. E. (1981). Growth of *Penicillium oxalicum* as a biological seed treatment on pea seeds in soil. *Phytopathology* **71,** 929–933.

Woodruff, H. B. (1980). Natural products from microorganisms. *Science, N.Y.* **208,** 1225–1230.

Wright, J. M. (1956). Biological control of a soil-borne *Pythium* infection by seed inoculation. *Pl. Soil* **8,** 132–140.

Wright, V. F., de las Casas, E., and Harein, P. K. (1980a). The nutritional value and toxicity of *Penicillium* isolates from *Tribolium confusum*. *Env. Ent.* **9,** 204–212.

Wright, V. F., de las Casas, E., and Harein, P. K. (1980b). Evaluation of *Penicillium* mycotoxins for activity in stored product Coleoptera. *Env. Ent.* **9,** 217–221.

Wright, V. F., Harin, P. K., and Collins, N. A. (1980c). Preference of the confused flour beetle for certain *Penicillium* isolates. *Env. Ent.* **9,** 213–216.

Wright, V. F., Vesonder, R. F., and Ciegler. A. (1982). Mycotoxins and other fungal metabolites as insecticides. In "Microbial and Viral Pesticides" (E. Kurstak, ed.), pp. 559–583. Marcel Dekker, New York and Basel.

9 Ambrosia Galls: The significance of fungal nutrition in the evolution of the Cecidomyiidae (Diptera)

J. BISSETT AND A. BORKENT

Biosystematics Research Centre, Agriculture Canada, Ottawa, Canada

Abstract

A review of the literature on fungi inhabiting galls of Cecidomyiidae provides evidence that gall midge larvae in the Lasiopterini and Asphondyliidi have specific fungal symbionts. These fungi are referable to *Macrophoma* or related coelomycete genera, and are probably anamorphs of *Botryosphaeria* (Ascomycotina). The association between fungus and insect in these "ambrosia galls" is apparently mutualistic. The gall midge larva is dependent on the mycelium of the fungus as a food source, and the fungus relies on the female adult midge for transportation and inoculation onto the plant host. Morphological adaptations to transport conidia of *Macrophoma* occur in adult females of Lasiopterini and Asphondyliidi, though the conidia-carrying structures are independently derived in these two taxa and not homologous. Cladistic data indicate that the endomycetophagous mode of nutrition is plesiotypic in the subfamily of plant gall-forming cecids (Cecidomyiinae). The endomycetophagous nutrition of the ambrosia gall midge larvae may facilitate shifts to new hosts (occurring during the adult stage), and the association with fungi at the origin of the gall-forming habit could explain the rapid radiation of all the gall-forming cecids across phylogenetically diverse plant taxa. Since some species of *Botryosphaeria* and related anamorphs cause canker-like galls on vascular plants, the original ability to induce galls may have been an attribute of the fungal symbionts, though the cecid larva eventually acquired or evolved a gall-inducing capability.

INTRODUCTION

The Cecidomyiidae is a diverse and remarkably speciose family of Diptera, with over 3000 described species. They are popularly known as gall midges,

COEVOLUTION OF FUNGI
WITH PLANTS AND ANIMALS
ISBN 0-12-557365-0

even though the gall-forming habit is restricted to only a few lineages of one of the three subfamilies. It is chiefly through marked radiation of these few lineages that the gall formers have become such a visible component of many communities. Although many species of cecids are the sole occupants of their galls, some kinds of galls are cohabited by specific fungi that provide a food source for the developing larva. Neger (1909) named these "ambrosia galls", recognizing the interdependence of the fungus and insect in these associations.

A moderate amount of literature has accumulated describing and speculating on the nature and significance of the various feeding modes found in the Cecidomyiidae. Larvae of cecids may be free-living fungal feeders, gall formers, free-living plant feeders, predators, parasites, or as noted above, fungal-feeding in ambrosia galls. As a result of such diversity, a variety of evolutionary scenarios have been proposed to interpret the origins of these strategies.

All plant feeding cecids are in the subfamily Cecidomyiinae. Some tribes in this subfamily are predominantly free-living fungus feeders, with a few representatives that are parasitic or predatory. Phytophagy is represented by larvae that are free-living in flower heads and plant buds, and by a very large and diverse assemblage of gall formers, the latter restricted to the tribes Oligotrophini, Dasineurini, Clinodiplosini and Cecidomyiini. The other cecid gall formers, in the Lasiopterini and Asphondyliidi, produce ambrosia galls in which the larval chamber is at some stage of development lined with the mycelium of a specific fungus. Although these taxa also have generally been considered phytophagous (e.g., Roskam, 1985), there is increasing evidence that the nutritive tissues in the gall are derived from the mycelium of the fungal associate rather than from the plant tissues, and the larval mode of nutrition is probably mycetophagous. The development of this association is thought to have occurred separately in the Lasiopterini and Asphondyliidi. Some modern authors consider the ambrosia gall-forming cecids to have reverted to the mycetophagous mode of nutrition from phytophagous gall-forming ancestors (e.g., Gagné, 1986).

Here we provide an overview of what is known about ambrosia galls and previous interpretations of the significance of the relationship between these midges and their fungal associates. Finally, we discuss the consequences of our present knowledge of ambrosia galls on our understanding of the evolution of feeding modes in the Cecidomyiidae.

THE DEBATE ON THE ECOLOGICAL SIGNIFICANCE OF AMBROSIA GALLS

Trelease (1884) examined the leaf blister galls caused by *Asteromyia* species, which are widely distributed in North America on *Solidago* and *Aster*, and was the first to conclude that these galls were a product of the activities of the insect larva and fungus which they contained. Although the two kinds of organisms always occurred together in these galls, Trelease postulated that the insect paved the way for the fungus, with the hyphae of the fungus unable to penetrate the uninjured plant.

Baccarini (1893) first alluded to a mutualistic symbiosis between a specific fungus and a gall midge larva (probably *Asphondylia capparidis*) in flower galls on *Capparis spinosa*, and he proposed the term "mycozoocecidium" for these kinds of galls, concluding that both fungus and insect contributed to the develoment of the gall. Trotter (1900) observed a similar association between *Asphondylia* and a fungus, and later (Trotter, 1905) also advocated the symbiotic theory for the associations described by Trelease and Baccarini. Neger (1909, 1910) was a proponent of the mutualistic nature of these gall associations, and proposed the name "Ambrosiagallen" (ambrosia galls) for them, an analogy with the better known mutualistic association involving some wood-inhabiting scolytid beetles that cultivate specific fungi ("ambrosia fungi") in the tunnels in which brood is reared. Neger (1910) examined the galls of *Asphondylia genistae* and *A. mayeri* on buds and fruit respectively of *Sarothamnus scoparius*, and of *Asphondylia coronillae* on buds of *Coronilla emerus*. He found a slightly developed mycelium at the side of the newly hatched larva and assumed that the female gall midge collected spores from pycnidia that fruited on the surface of the gall at the time of adult emergence. Neger noted that growth of the larva was correlated with the production of mycelium in the gall chamber. However, he concluded that the insect larva was primarily responsible for differentiation of the gall tissues, leading him to reject the term mycozoocecidia proposed by Baccarini for these kinds of galls.

Docters van Leeuwen (1929, 1939) described the gall on *Symplocos fasciculata* caused by *Asphondylia bursaria* as a symbiotic association, with the larva initiating gall development, and later relying on the mycelium of the fungal associate as the sole food source. He observed conidia of the fungus inside the gall chamber immediately following oviposition by the female. Although he concluded that the female midge placed the conidia in the gall chamber, he was unable to propose a mechanism by which this might have been accomplished. For the gall caused by *Lasioptera carophila* on axes of inflorescenses of *Pimpinella saxifraga*, Docters van Leeuwen (1939) reported that the plant component of the chamber wall consisted mainly of lignified cells filled with air, from which the larvae would be unlikely to derive any nutrition. He concluded that the larvae were nourished instead by the fungi developing on the wall of the chamber. Buchner (1930) concurred on the symbiosis of the two elements, noting for example that the flower gall of *Asphondylia echii* on

Echium vulgare consistently contains fungi, whereas the flower gall on the same host caused by *Contarinia echii* does not.

Ross (1932), however, argued against a mutualistic relationship between cecid larvae and fungi, concluding that the larvae carry the fungus to the place where the gall occurs incidentally. He believed that the fungus lives saprophytically as an inquiline in the gall chamber, with competition resulting between the larva and the fungus. Ross (1932) employed the term "verpiltze Mückengallen" (mouldy midge galls) for these associations. His conclusions were based, in part, on his failure to isolate consistently the fungal associate from the galls. He also concluded that eggs of *Asphondylia* were deposited on the outside of floral buds, incorrectly believing that the ovipositor of the female was too soft and short to penetrate the host, and that the larvae on entering the host tissues drag inside any fungal conidia that happen to be present. By this time, the occurrence of fungi in some cecid galls was known to be widespread, and Ross (1932) compiled the known examples, all associated with galls of *Asphondylia*, *Schizomyia*, and *Lasioptera*.

Puzanowa-Malysheva (1935) and Goidanich (1941) provided detailed accounts of the development of the galls produced by *Asphondylia prunorum* on buds of *Prunus* spp. Both authors concluded that the relationship was mutualistic for the insect and fungus. Goidanich successfully isolated and identified the fungal associate, and noted that most fungi identified in the various gall associations studied to that time appeared to be taxonomically related. Meyer (1952) also provided evidence for the symbiotic theory of Neger, observing that the mycelium lining the gall chamber was essential to the nourishment of the larvae of *Lasioptera eryngii* on *Eryngium campestre*, and of *L. carophila* on *Pimpinella saxifraga*.

Batra (1964) performed a more extensive study of the development of the blister galls on Astereae caused by *Asteromyia* species, which were earlier studied by Trelease (1884). Batra recognized that the galls were the result of combined activities of fungus and larva. However, he concluded that the insects in the blister galls were associated with leaves of *Solidago* only after attack by the fungus. Batra did not rule out the possibility of mutual benefit from this association, concluding that the fungus might utilize larval frass as a supplementary source of nutrients and in return provide shelter to the insect. Nevertheless, Batra and Lichtwardt (1963) considered the term "ambrosia galls" to be misleading for cecid and other kinds of plant galls with fungal associates. They examined a wide range of insect galls containing fungi, including some cecid galls, and classified the associations into two kinds: (1) apparently specific fungal associations with "sap-sucking" insects like *Asphondylia* and *Lasioptera*, in which the mycelium of the fungus is apparent during the early stages of gall formation and eventually lines the inner surface of the gall in a palisade-like layer, and (2) non-specific associations often involving more than one species of fungi that grow sparingly and randomly in the gall chamber. Their objection to the use of the term "ambrosia gall" was based on their conclusions that the insect and fungus are capable of living independently in all of these associations; and that the gall fungi are taxonomically diverse and not species-specific, and are airborne rather than borne by the insect associate.

Borkent and Bissett (1985) discovered morphological adaptations to transport conidia, analogous to the mycangia in ambrosia beetles, in adult females of the ambrosia gall midges. These conidiophorous structures were not homologous in the Asphondyliidi and Lasiopterini, indicating that the mycangia evolved independently in these two groups. The discovery that conidia of the fungal associate were purposefully transported by the insect associate in the ambrosia galls disproved the final remaining objection to Neger's (1910) symbiotic theory. Consequently, the term "ambrosia galls" is appropriate for the galls induced by Asphondyliidi and Lasiopterini and characterized by "endomycetophagous" larval feeding.

THE ANATOMY AND DEVELOPMENT OF AMBROSIA GALLS

Most cecid galls can be incited only in undifferentiated meristematic tissue. Typically they develop by the inhibition of normal cell differentiation of the plant tissues proximal to the larva or larval chamber, followed by abnormal cell elongation, proliferation, and redifferentiation in the tissues forming the walls of the gall. The gall tissues characteristically retain a juvenile state and often continue to grow after normal tissues surrounding the gall have ceased growth. Secretions from gall midge larvae apparently cause the physiological events associated with gall differentiation, although the nature of these secretions and their regulatory role is not understood. A similar sequence of events occurs in most of the ambrosia galls that have been examined to date. The gall differentiates in response to the secretions or activities of the gall midge larva, with the fungal associate in most cases apparently making little or no contribution to the development of the gall.

In galls of endophytophagous gall midges, cytoplasmically dense nutritive cells develop from plant tissues lining the gall chamber in response to larval feeding. The nutritive tissues are supplied by a proliferation of vascular bundles that surround the gall chamber. In ambrosia galls, however, nutritive cells do not develop from the tissues of the plant. Instead, the larva is dependent on the mycelium that develops inside the larval chamber as its only apparent food source. Meyer and Maresquelle (1983), studying the floral gall

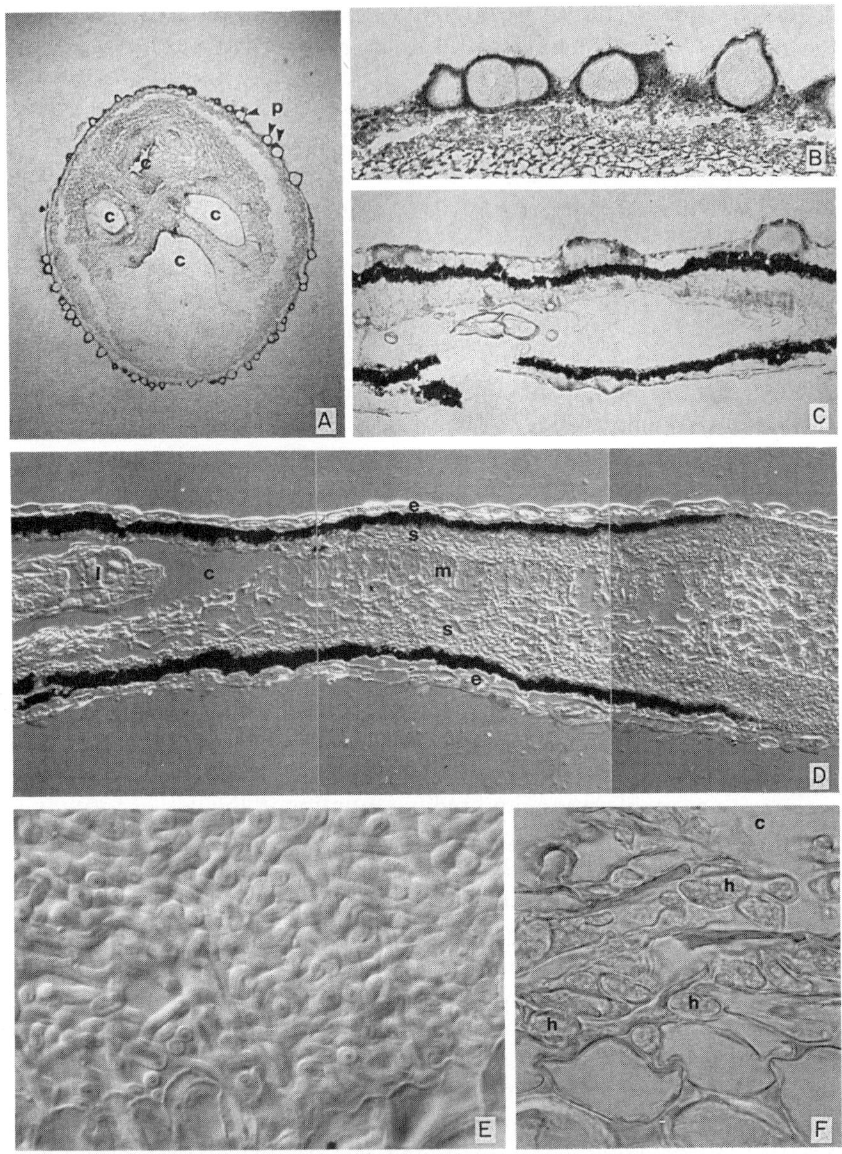

Figure 1. Cecid galls and associated fungi. **A,** Cross section through the fruit gall on *Ilex opaca* caused by *Asphondylia ilicicola* (×7); the larval chambers (c) are surrounded by palisade-like plant tissues; pycnidia (p) of *Dothiorella ilicella* fruit on the outer surface. **B,** Pycnidia of *D. ilicella* (longitudinal section) on *Ilex opaca* fruit gall (×45). **C,** Pycnidia of *Macrophoma gallicola* (longitudinal section) on leaf blister galls on *Solidago mollis* caused by *Asteromyia carbonifera*, ex isotype of *M. gallicola*,

of *Kiefferia pimpinellae* on *Pastinaca silvestris*, reported that the mycelium of a fungus growing intracellularly in the wall of the gall formed ampoules. These ampoules, located on free ends of the mycelium extending into the larval cavity, may function in larval nutrition.

There are several detailed descriptions of gall development for species of *Asphondylia* (Docters van Leeuwen, 1929, 1939; Goidanich, 1941; Freeman and Geoghagen, 1987). These descriptions, and our unpublished observations on *Asphondylia ilicicola* causing a fruit gall on *Ilex opaca* (Figure 1A, B), and on *A. psilostachys* on stems of *Ambrosia* (Figure 1F), indicate a relatively uniform morphology and pattern of development for galls of the Asphondyliini. The female midge pierces the plant tissues with her ovipositor, creating a small chamber, inside which are deposited an egg and a few conidia. The egg hatches within a few days. However, the development of the fungus may be suspended for many weeks, leading to the conclusion that gall formation is initiated by the larva. Normal cell differentiation is arrested, and a fast-growing meristematic tissue is formed around the gall chamber. The tissues surrounding the gall chamber remain more or less embryonic, and often there is increased vascularization around the gall chamber. After inducing gall formation, the larva may not develop for a time, although the gall and gall chamber increase rapidly in size. After gall development is well established, the fungus begins to develop rapidly, eventually forming a compact tissue that nearly fills the gall chamber, though initially not significantly penetrating the tissues of the gall itself. A small cavity within the gall, where the larva remains undeveloped, may remain free of mycelium. Finally the larva begins to grow rapidly, consuming the mycelium, which appears to be its only food source.

Gall development in the Lasiopterini has not been examined as carefully, although the limited data at hand indicate that the sequence of events in gall formation in this tribe may be relatively variable. Meyer (1952) observed stimulated cambial development in the galls of *Lasioptera rubi* on twigs of *Rubus*. Nourishing tissues were not differentiated from the host plant. Instead, the gall cavity was filled with a mycelium that also grew intracellularly on the inner surface of the larval chamber in the early stages of the develop-

---

DAOM 81755 (×45). **D,** Vertical section through the leaf blister gall on *Solidago caesia* caused by *Asteromyia* (×180); prosenchymal fungal stroma (s) occupies a zone between the leaf mesophyll (m) and epidermis (e); the stroma is absent or reduced in the larval chamber (c), and has presumably been consumed by the developing larva (l). **E,** Prosenchymal fungal stroma in blister gall on *Solidago canadensis* (×450). **F,** Swollen nutritive hyphae (h) proximal to larval chamber (c) in the stem gall on *Ambrosia* caused by *Asphondylia psilostachys* (×450).

ment of the gall. Meyer also observed the absence of nutritive tissues and appearance of mycelium in the larval chamber of galls of *Lasioptera eryngii* on *Eryngium campestre*. He concluded that the larva was dependent for nourishment on the mycelium of the fungus introduced by the gall maker. Our observations on a stem gall on *Baccharis halimifolia* caused by *Neolasioptera lathami* were similar, although the mycelium of the fungal associate was less extensive and did not fill the larval chamber as a compact tissue, which appears to be the usual condition in the Asphondyliini.

We have also studied the gall caused by an undescribed *Astictoneura*, formed at the base of the stem of a *Muhlenbergia* species near soil level. Free-flying females collected in the vicinity carried conidia of a single *Macrophoma* species. We have not studied the gall in its early stages of development, but during latter stages the gall tissues are extensively invaded by the mycelium, stromata and conidiogenous structures of a variety of fungi. Although the female gall midge introduced a single species of fungus at the time of oviposition, in the later stages of larval development it is clear that the tissues of the gall are invaded by a number of other fungi, probably resident in the soil in which the base of the gall is situated. The possibility should be considered that these fungi contribute to the nutrition of the insect.

The "blister galls" on leaves and stems of Astereae caused by species of *Asteromyia* may have an exceptional form of development. The larva of *Asteromyia* does not induce hyperplasia and hypertrophy in the plant tissues comprising the gall, and the fungal associate is primarily responsible for the gall morphology that results from rapid proliferation of the fungus intracellularly in the plant tissues around the larval chamber (Camp, 1981). According to Gagné (1968, 1986), the larvae of *Asteromyia* appear stationary and are appressed to the leaf veins. The leaf tissues remain green in the immediate vicinity of the larva, appearing succulent until the larva is fully grown. Gagné suggested that the larvae feed on plant tissue. He did not propose a role for the fungus in these galls. However, our observations on *A. carbonifera* and *A. euthamniae* indicate that the larvae are mobile during their development and may be nurtured by the mycelium that fills the larval chamber (Figure 1D, E).

Gagné (1986) concluded that many Lasiopterini appear not to feed on the fungus growing in their galls, and that some species may not have fungal associates. However, fungal associates have been noted in every carefully studied Lasiopterini gall association reported in the literature (e.g. Bronner, 1977), and also in those associations studied by us. In some instances the growth of the fungal symbiont may be inconspicuous and largely confined to the inner surface of the gall chamber. Furthermore, mycetophagy appears to be the significant mode of nutrition in every study in which larval feeding has been examined.

IDENTITY AND BIOLOGY OF THE FUNGAL ASSOCIATE

Attempts to identify the fungal associate in ambrosia galls have frequently been unsuccessful. Very often the fungus does not fruit on the host plant while it is inhabited by the insect associate (e.g., Batra, 1964), and frequently the fungal associate does not grow or sporulate in culture. Isolations from the gall interior and from the larva more often than not have resulted in the isolation of non-specific fungi or contaminants (e.g., Baccarini, 1893; Ross, 1932; Batra and Lichtwardt, 1963).

Schweinitz (1822, 1832) provided three names in *Rhytisma* for the fungi occurring in galls of *Asteromyia* on Astereae, distinguishing the species by the external features of the galls. The fungus did not fruit on the galls, and these forms were assigned to *Rhytisma* on the basis of the superficial resemblance of the galls to the tar-spot infections produced by that genus of ascomycetes. Batra (1964) interpreted the galls as sclerotia and transferred all three species described by Schweinitz to *Sclerotium asteris.* He successfully isolated the specific fungus from the "blister galls" on Astereae, but could not induce the fungus to sporulate in culture. This fungus was identified by Borkent and Bissett (1985), who noted that conidia carried by the female gall midge were identical to *Macrophoma gallicola*, originally described from leaf blister galls on *Solidago mollis* (Figure 1C). We have subsequently induced this fungus to fruit on galls of *Solidago* species incubated at 2–3°C in the laboratory, and also once found the fungus fruiting in the spring on a galled leaf of *Solidago* that had fallen to the ground the previous autumn.

Neger (1910) observed pycnidia of *Macrophoma* associated with galls of *Asphondylia*, identifying the fungus on the galls of *A. genistae* and *A. mayeri* as *Macrophoma coronillae*, and giving the name *M. coronillae-emeri* to the fungus on the galls caused by *A. coronillae*. Goidanich (1941) identified the fungal associate in the galls of *Asphondylia prunorum* as a species of *Sphaeropsis*. The fungus fruits on the exterior surface of galls on *Prunus* at the time of emergence of the adult midges. The fungi described by Neger (1909) and by Goidanich (1941) closely resemble the *Dothiorella* anamorph of *Botryosphaeria ribis*. Goidanich discussed the variability of characters distinguishing the genera *Macrophoma*, *Sphaeropsis*, and *Diplodia*, noting the probable biological affinity of forms ascribed to these three genera. Taxonomically related fungal associates were reported by Batra and Lichtwardt (1963) for *Asphondylia cytisii* on *Cytisus praece* bud galls, and by Kaiser (1978) from the gall caused by *Lasioptera rubi*.

The fungus associated with *Asphondylia ilicicola* on fruit galls of *Ilex opaca* forms pycnidia on the surface of the gall at the time of emergence of the adult gall midge (Figure 1B). We have identified this fungus as *Dothiorella ilicella* (syn. *Macrophoma ilicella*). This fungus is hardly distinguishable from the anamorph of *Botryosphaeria ribis*, which is widely distributed on diverse hosts, and for which the anamorph has probably been redescribed many times from the different hosts. We have also successfully isolated a mycelium resembling *Botryosphaeria* in culture from the galled tissues and larva of *Asphondylia psilostachys* on an *Ambrosia* species, although this has not sporulated.

Further evidence that the fungal associates in ambrosia galls are phylogenetically closely related comes from observations on the morphology of conidia contained in the mycangia of the female gall midges. Figure 2A–D illustrates a range of conidia from mycangia of taxonomically and geographically diverse Asphondyliidi, and Figure 2E–H those from Lasiopterini. All these conidia share the following characteristics: they have a similar shape and vary primarily in length; the base is truncate and indicative of holoblastic ontogeny; they are hyaline and aseptate or tardily develop brownish pigments and one, or rarely two, septa. All may be referred to one or more of the following closely related and poorly differentiated coelomycete anamorph genera: *Diplodia*, *Dothiorella* (?=*Botryodiplodia*), *Fusicoccum*, *Macrophomopsis*, *Sphaeropsis* (?=*Macrophoma*). Excluding the numerous species that should be transferred to unrelated genera, all known teleomorph connections for these anamorph genera are in *Botryosphaeria* (Ascomycotina, Dothideales, Botryosphaeriaceae). Arguably, all of the anamorphs observed in the gall midge mycangia could be considered congeneric. However, the determination of the correct generic name(s) for these requires taxonomic investigations beyond the scope of the present paper. In addition, species identifications in this group are to a large degree not yet possible. For the purposes of the discussion that follows, conidia of fungi associated with ambrosia galls are referred to as *Macrophoma* (Borkent and Bissett, 1985).

There is a profusion of names for the anamorphs. *Diplodia*, *Sphaeropsis*, and *Macrophoma* together have over 2000 epithets assigned to them. Some of these should be referred to unrelated genera, but the majority are undoubtedly synonyms. The actual number of species of *Botryosphaeria* and related anamorphs will not be known until the extensive synonymy is studied. There may be fewer species of fungi than cecids involved in the ambrosia gall associations (i.e., several cecid species may use the same species of fungus).

Species of *Botryosphaeria* are saprophytes or weak parasites occurring over a phylogenetically broad range of gymnosperms and angiosperms. The teleomorph usually matures on dead woody tissues. A few species are relatively well known; several cause fruit rots on a variety of angiosperms, and *Botryosphaeria dothidea* and *B. obtusa*, for example, cause a variety of diseases over a wide range of mostly woody hosts. The anamorphs are also found on a wide range of hosts. The anamorph fruits more readily than the teleomorph, on woody and herbaceous substrata, and occasionally also on

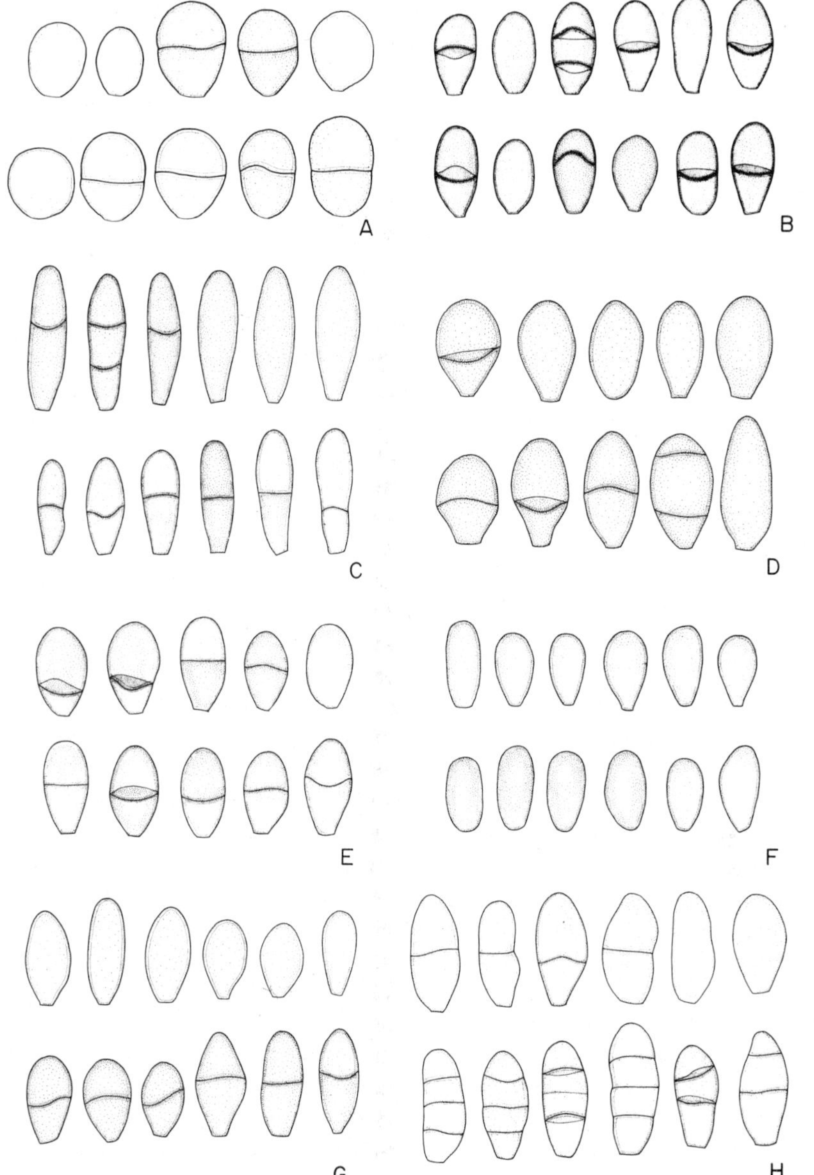

Figure 2. *Macrophoma* conidia contained in the mycangia of various ambrosia gall midges (×1000). **A,** *Asphondylia* (Entebbe, Uganda). **B,** *Asphondylia* (Catamara, Argentina). **C,** *Asphondylia* (New Brunswick, Canada). **D,** *Schizomyia* (Chiapas, Mexico). **E,** *Lasioptera* (Deganya, Israel). **F,** *Ozirhincus* (Ontario, Canada). **G,** *Asteromyia* (Ontario, Canada). **H,** *Neolasioptera* (Ohio, USA).

infections on living host tissues. Only the anamorphs have been observed to fruit on the ambrosia galls.

Parasitism in *Botryosphaeria* and related anamorphs is usually necrotrophic. However, in the ambrosia gall associations, the mycelium is intracellular and nutrition appears to be biotrophic, that is, derived from living host cells. Some species and anamorphs of *Botryosphaeria* that do not have insect associates produce gall-like cankers on a variety of living plants, but their nutritional relationship with the host plant has not been studied.

THE MYCANGIA OF AMBROSIA GALL MIDGES AND CONIDIAL TRANSPORT

Docters van Leeuwen (1929) first reported that the female midge of *Asphondylia bursaria* deposited 1–4 brownish conidia inside the egg chamber in the plant tissue during oviposition. He calculated that each female would have to carry perhaps 200–300 conidia as inoculum for 80–100 eggs, but he failed to observe any conidia carried by the female midge. Goidanich (1941) also observed that conidia of the fungal associate were deposited alongside each egg of *Asphondylia prunorum* inside the host tissues. Although both authors concluded that the female was responsible for the conidia occurring inside larval chambers, they were unable to propose a mechanism by which this was accomplished. Goidanich observed that the fungus fruited on the outer surface of the old gall at the time of emergence of the adult, and postulated that the conidia might be borne among the hairs on the legs of the female.

Borkent and Bissett (1985) observed that female adults of ambrosia gall midges possess specialized conidia-carrying structures ("mycangia", Batra, 1963) that ensure inoculation of a specific fungus onto the host plant during oviposition. In most Alycaulina (tribe Lasiopterini) a pair of dorsolateral, elongate pockets on abdominal segment 9 comprise the mycangia (Figure 3A). When the ovipositor is retracted, segments 8 and 9 including the mycangia are contained by segment 7, with the terminal cercus externally visible at the apex of segment 9. This type of mycangium was seen in *Asteromyia*, *Neolasioptera*, *Calamomyia*, *Chilophaga*, *Astictoneura*, and *Edestochilus*. Mycangia may be lacking in *Protaplonyx* and *Meunieriella*.

Lasiopterina also have a protrusible ovipositor similar to that in Alycaulina. They have two dorsilateral groups of strongly developed setae situated at about one-third the length of segment 8 and similar setae on the cercus (Figure 3B, C). When the ovipositor is retracted, each set of setae on segment 8 folds in on itself to form a lateral pocket in which conidia are entrapped. The well-developed setae on the cercus also entrap many conidia.

Females of Asphondyliidi have needle-like ovipositors (modified cerci)

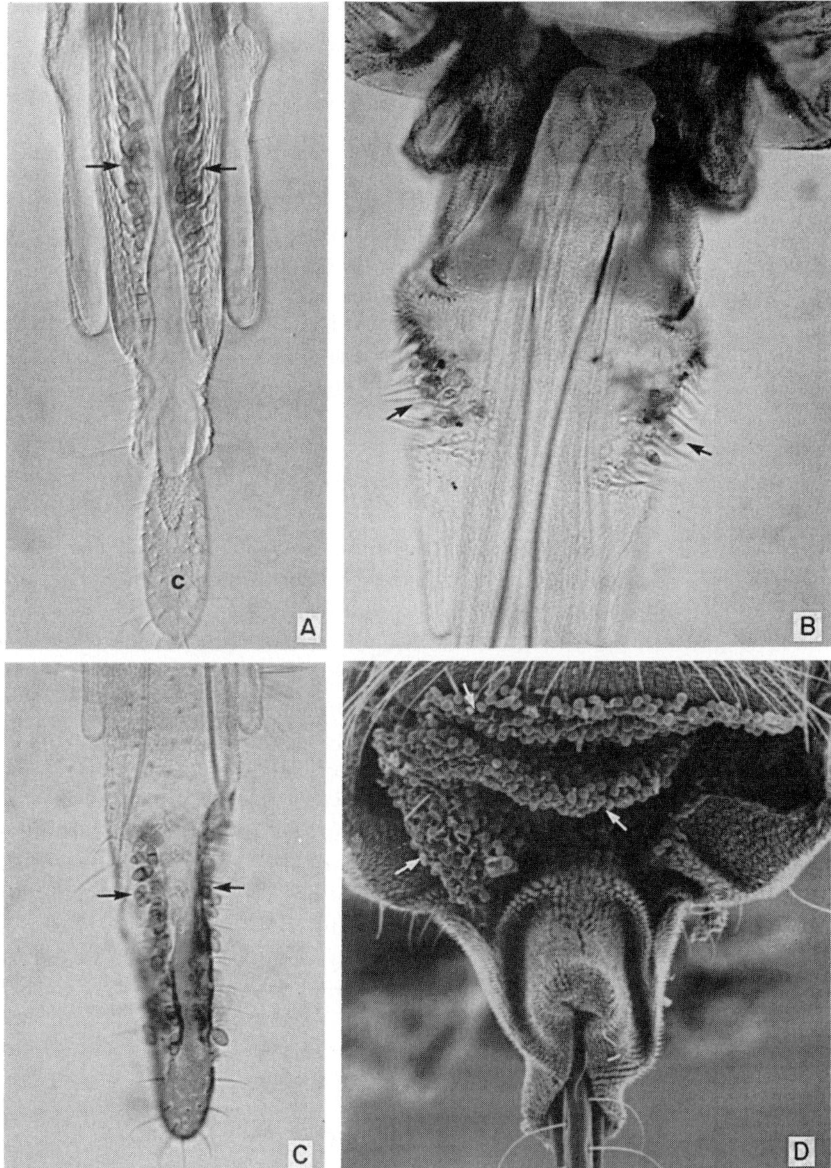

Figure 3. Mycangia of Cecidomyiinae. **A,** *Calamomyia* in dorsal view (× 240) showing mycangia (arrows) on abdominal segment 9, cercus (c). **B,** *Lasioptera* in dorsal view (× 240) showing setae surrounding mycangia (arrows) on abdominal segment 8. **C,** As (**B**) but on the cercus. **D,** *Schizomyia* in ventral view (× 150) showing conidia (arrows) borne in the large membraneous mycangium dorsal to abdominal sternite 7.

that are ensheathed within the abdomen when retracted. Conidia are carried in a comparatively large membranous mycangium positioned dorsal to the posterior portion of abdominal sternite 7, extending laterally nearly to the margins of sternite 7 and opening along the posterior margin of the sternite (Figure 3D).

Borkent and Bissett (1985) concluded that the mycangia of the Lasiopterini and of the Asphondyliidi are not homologous and are therefore independently derived. The relationships within the Lasiopterini were insufficiently resolved to determine whether the mycangia of Alycaulina and Lasiopterina are also independently derived.

Conidia in the mycangia of Alycaulina and Lasiopterina are nearly always of the *Macrophoma* kind. A total of about 40–150 conidia are carried in the paired mycangia of the adult female in these subtribes. Usually higher numbers of conidia are carried in the comparatively large mycangium of female Asphondyliidi. Occasionally, a few conidia of other kinds of fungi, setae and miscellaneous detritus are also present in the mycangium of Asphondyliidi.

Nearly all field-collected specimens we have examined were carrying conidia and nearly all reared specimens were not. In most ambrosia gall midges we have studied, the fungal associate does not fruit on the galls or other parts of infected plants at the time the adult midges are emerging. This may explain why previous workers have had little success in attempts to make reared females induce galls in proffered plants (Gagné, 1968; Weis *et al.*, 1983). Females must collect conidia from some other location, possibly from old galls or other components of the ground litter, or even another host plant species. The larvae of some Asphondyliini leave the gall to pupate in the soil. On emergence the female might find conidia in the litter, or by returning to the old galls if these are still attached to the plant. The ubiquitous presence of ambrosia gall midges geographically, ecologically, and chronologically suggests that conidia of *Macrophoma* are nearly universally available. However, we have never succeeded in observing how the female midge finds and collects the conidia of what has previously been regarded as a relatively uncommon fungus.

THE ADAPTIVE NATURE OF GALL FORMATION

The major hypotheses on the adaptive significance of gall formation were reviewed by Price *et al.* (1986, 1987). The adaptive value of the gall-forming habit for insects is well supported. The gall provides improved nutrition for the insect, evidenced by higher concentrations of nutritive compounds in gall tissues and decreased chemical defences of the plant. The microenvironment is also improved for the insect, with protection against hygrothermal stress. Galls may provide increased protection against predation, although protection against parasitoids is less evident. Mani (1964) proposed that galls have an adaptive value for plants also, as a form of defence where the plant encapsulates a herbivore. However, growth and reproduction appear to be reduced in galled plants similarly to plants that are subject to other forms of herbivory. Regardless, the presence of the gall association is clearly more of a reflection of cecidomyiid diversification than that of the plant hosts. The development of galls is restricted to only a few lineages of Cecidomyiinae but covers a wide array of unrelated plant families. There is no evidence for the radiation of plant taxa having the potential for gall formation. This indicates that it was the gall-formers that radiated across plant taxa, suggesting an adaptive advantage for the midges to do so.

As suggested by Skuhrava *et al.* (1984), fungi associated with ambrosia galls also may derive some adaptive benefit from the association. In all ambrosia gall associations studied to date, the fungus appears to be incapable of infecting the living tissues of the plant host outside of the gall association. Modification of plant tissues by cecid larvae may provide new host opportunities or improved nutrition for the fungus, or adult cecid behaviour may assist the fungus in dissemination or in overcoming host defences to infection. Fungi associated with ambrosia galls appear to benefit by being transported and inoculated onto the host by the insect. If the fungus is physiologically able to sporulate on the gall in synchrony with the emergence of the adult insect, then the dissemination of the fungus will be perpetuated by the emerging females. However, given the general absence of conidia carried by reared cecids, the fungus may benefit entirely by inoculation in advance of competing saprophytes onto a plant organ that is destined to become moribund, sporulating at a later date, and perhaps being collected and transported by subsequent generations of female cecids searching for conidia.

Finally, it is possible that there is no benefit to the fungus at all and that the cecids are using and transporting the fungus solely as a food source for their young. If so, this would draw into question the validity of considering the relationship between the fungus and midge as mutualistic.

COEVOLUTION BETWEEN GALL-FORMING CECIDS AND VASCULAR PLANT HOSTS

The family Cecidomyiidae is one of the youngest branches of the infraorder

Bibionomorpha (Wood and Borkent, in press). The presence of Cretaceous fossils at 74 Myr B.P. (Gagné, 1977) indicates that the family diversified at least to the level of the Heteropezini by that time. Most tribes of Cecidomyiinae were well represented by the upper Miocene (Gagné, 1973). The transition to phytophagy may have occurred at or about the late Cretaceous (Mamaev, 1968, 1975), with the marked radiation of the gall-forming Cecidomyiinae coinciding with the rapid expansion of angiosperms from that time.

Roskam (1985) demonstrated taxonomic interdependence between the diversification of a few gall midge species in the genus *Semudobia* and their hosts in *Betula*, showing that related gall-formers may live on related hosts. Some cecid taxa are restricted to certain groups of vascular plant hosts (Roskam, 1985). For example, all species of *Ozirhincus* induce fruit galls on Antherideae (Asteraceae), those of *Asteromyia* only in leaves and stems of Astereae (Asteraceae), and those of *Asphondylia* mostly on Fabaceae or Labiatae. However, it is uncertain whether coevolution has occurred in these cases. Phylogenetic studies are required to examine the possibilities of congruent phylogenies. Seemingly incongruent phylogenies are by far the most common pattern between the gall midges and their vascular plant hosts, apparently due to shifts of gall midges to unrelated host species during the free-living adult phase. In general, ecological opportunists shifting to new hosts in proximity, and not necessarily to those phylogenetically related to the original hosts, appear to be comparatively common among the Cecidomyiidae.

CURRENT EVOLUTIONARY SCENARIOS INTERPRETING FEEDING MODES IN THE CECIDOMYIIDAE

There is general agreement that the plesiotypic feeding mode of the Cecidomyiidae is free-living fungal feeding (Gagné, 1986; Mamaev, 1964; Roskam, 1985), as evidenced by outgroup comparisons discussed below. The transition to endophytophagy is generally considered to have been polyphyletic (Gagné, 1986; Mamaev, 1964; Roskam, 1985), evidently occurring independently in the Lasiopteridi and Cecidomyiidi, since both supertribes have forms that are free-living mycetophages. A third transition to phytophagy was argued by Gagné (1986) to have occurred in the Clinodiplosini, which has free-living fungus- and plant-feeders as well as gall-forming members. The evolution of the gall-forming habit in cecids was proposed to have occurred through the transition of sedentary phytophages to endophytophagy by production of differential plant growth (e.g., Gagné, 1986; Price *et al.*, 1986).

Gagné (1986) has suggested that the Asphonyliidi and the Lasiopterini have secondarily regained an association with fungi.

A NEW MODEL OF THE EVOLUTION OF GALL-FORMING BEHAVIOUR IN THE CECIDOMYIINAE

The divergence of ecological interactions is best interpreted in the context of an understanding of phylogenetic relationships (Anderson, 1982; Morse and White, 1979). Clearly argued cladistic evidence for relationships has generally not been provided for the Cecidomyiidae. Although our understanding of these relationships among the Cecidomyiidae is scanty, some relationships are known and can be used to hypothesize transformation series for the mode of feeding. However, further phylogenetic studies are required to test the ideas proposed here.

The following discussion is based on the phylogenies published by Gagné (1976) and Roskam (1985). The cladograms published by Gagné (1986) that show further resolution were not supported by argued synapomorphies and have not been incorporated into our analysis. Relationships between the Cecidomyiidae and other Bibionomorpha are discussed by Wood and Borkent (in press). Roskam (1985) has provided an excellent discussion of the host plant associations in the Cecidomyiinae.

As indicated by outgroup comparisons, it is clear that the plesiotypic feeding mode in the Cecidomyiinae is mycetophagy (Figure 4). This is supported by the presence of fungal feeding in the closest relatives of the Cecidomyiidae: the Sciaridae, and Mycetophilidae, and in the basal lineages of the Cecidomyiidae: the Lestremiinae, and Porricondylinae, both of which are probably paraphyletic in relation to higher taxa. In addition, some lineages within the Cecidomyiinae also are free-living fungal feeders.

Within the Cecidomyiinae, three main lineages diverge: the Cecidomyiidi, Asphondyliidi, and the Lasiopteridi (Figure 4). Synapomorphies that might further resolve relationships among these lineages require testing by outgroup comparisons. The Lasiopteridi divide into a lineage composed of the Ledomyiini, Brachineurini, and Rhizomiini (probably paraphyletic), and a lineage comprising the Oligotrophini–Dasyneurini (possibly paraphyletic) plus the Lasiopterini. The Alycaulina and Lasiopterina are sister groups within the Lasiopterini.

We have indicated in Figure 4 the larval feeding modes of each of the main lineages of Cecidomyiinae. Because of the outgroup comparisons given above and the presence of fungal feeding in the Ledomyiini, Brachineurini, and Rhizomiini, and in some members of the Cecidomyiini, an apparent

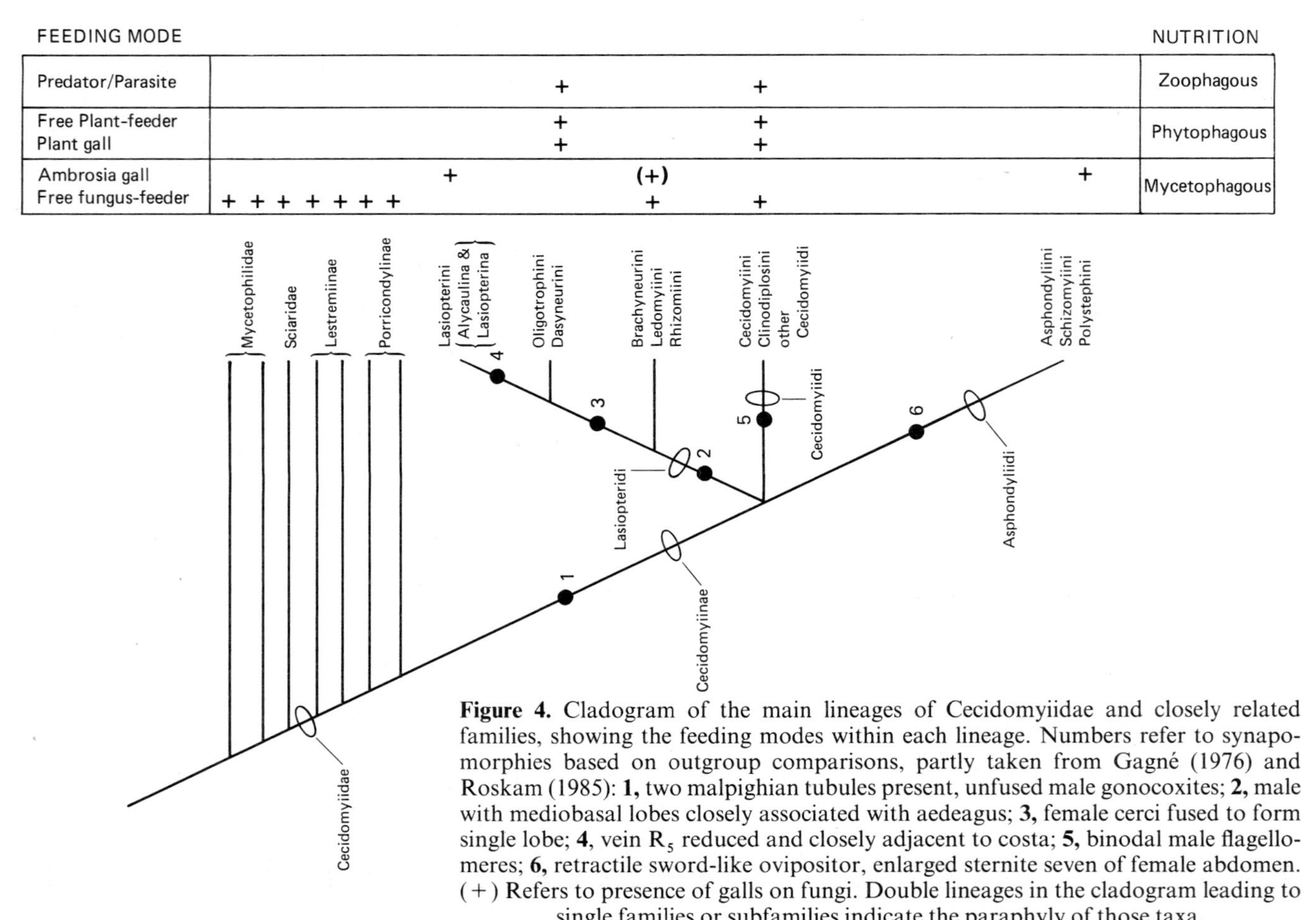

Figure 4. Cladogram of the main lineages of Cecidomyiidae and closely related families, showing the feeding modes within each lineage. Numbers refer to synapomorphies based on outgroup comparisons, partly taken from Gagné (1976) and Roskam (1985): **1,** two malpighian tubules present, unfused male gonocoxites; **2,** male with mediobasal lobes closely associated with aedeagus; **3,** female cerci fused to form single lobe; **4**, vein R_5 reduced and closely adjacent to costa; **5,** binodal male flagellomeres; **6,** retractile sword-like ovipositor, enlarged sternite seven of female abdomen. (+) Refers to presence of galls on fungi. Double lineages in the cladogram leading to single families or subfamilies indicate the paraphyly of those taxa.

interpretation of the evolution of feeding modes is that the plant feeding cecids (including phytophagous gall formers and mycetophagous ambrosia gall inhabitants) have evolved independently from free-living fungal feeders. A consequence of this interpretation is that the evolution of gall-forming capacities evolved at least three times (in the Oligotrophini, Dasyneurini, plus Lasiopterini; in the Cecidomyiidi; in the Asphondyliidi).

It is interesting to note, however, that some members of the Ledomyiini have novel interactions with fungi. Instead of the freely moving larvae characteristic of the outgroup of the Cecidomyiinae, these larvae induce galls on the fungi they attack (Larew *et al.*, 1987). This suggests that they may have arisen from an ancestor with the capacity to form galls, and indicates that the endomycetophagous feeding mode is the plesiotypic condition in the Lasiopteridi. In addition, the Brachineurini, which are currently unplaced in the Cecidomyiinae, also are known to form galls in some fungi (Harris and Evans, 1979). The Rhizomiini, also presently unplaced, appear to be free-living fungal feeders.

These data suggest the following scenario: the appearance of the ambrosia gall is a plesiotypic feature of the Cecidomyiinae. In the original form of the ambrosia gall association, the female midge may have been attracted to a gall-like *Macrophoma* growth on a living plant (some extant *Macrophoma* species are still capable of producing gall-like cankers on plants). Initially the fungal associate may have been responsible for regulation of host cell functions and gall differentiation. In time the larval stage of the insect associate evolved or acquired this ability. Since cecid galls share some developmental similarities, it is likely that the ability of the larva to induce gall formation also evolved prior to the divergence of the Lasiopteridi, Asphondyliidi, and Cecidomyiidi.

In both the Lasiopteridi and Cecidomyiidi the disassociation of the fungus and insect may have occurred commonly, with some forms reverting to free-living fungal feeding and others proceeding to endophytophagy. Free-living phytophagy may be the most derived trophic level, evolving from endophytophages that radiated onto new hosts on which they were unable to overcome the host plant resistence to gall differentiation. The evolution of the predaceous and parasitic cecids is still conjectural. Further phylogenetic data are required to interpret the origins of free-living phytophagous, predaceous, and parasitic lineages.

Within the Lasiopteridi, the Ledomyiini had the capacity to form galls on some fungi and possibly some species reverted to free-living fungal feeding. Oligotrophini and Dasyneurini lost the fungal association and developed in galls free of the fungus; in some instances they became free-living plant feeders or predators. Within the Cecidomyiidi, some members of the Cecidomyiini reverted to free-living fungus feeding, some lost their fungal associations to become phytophagous gall formers and free-living plant feeders, and some became predators or parasites. It should be noted that, because the Cecidomyiini is presently recognized as monophyletic (Roskam, 1985), homoplasy in feeding mode must be recognized regardless of the evolutionary scenario. Either some members of the Cecidomyiini reverted to fungal feeding, or some members independently evolved the capacity to induce gall formation. The reversion to free-living fungal feeding from an ancestor that was an ambrosia gall former is an easier evolutionary step than the independent evolution of gall formation.

The proposal that the ambrosia gall formers represent the plesiotypic feeding mode in the Cecidomyiinae may explain the following observations.

(1) The Asphondyliidi and the Lasiopterini all have a symbiotic relationship with the same fungal taxon. The association of both groups with *Macrophoma*-like fungi (i.e., anamorphs of *Botryosphaeria*), and the presence of endomycetophagous feeding, support the idea that a common ancestor had evolved a similar association that originated before the divergence of the Lasiopteridi, Asphondyliidi, and Cecidomyiidi.

(2) Many cecid species have a narrow host spectrum, limited usually to a variable number of congeneric species. Polyphagous gall formers are relatively rare (Sylvén, 1979; Roskam, 1985). However, a significant number of the polyphagous species are ambrosia gall formers (e.g., Gagné and Hawkins, 1983). Indeed, some *Asphondylia* species alternate host plants during their life cycle (Orphanides, 1975; Yukawa *et al.*, 1983). The association with fungi may allow the ambrosia midge to radiate to taxonomically diverse hosts more readily, perhaps by fungal mediation in the manipulation of the host defence mechanism or as a result of the dependence on mycelial rather than plant nutriment. The gall-forming cecids overall have a comparatively broad host spectrum. Perhaps this could be attributed to endomycetophagous nutrition being more prevalent than previously realized in the initial evolution of the gall-forming habit in this group. The association of a fungus at the origin of the gall-forming habit may explain the rapid radiation of all gall-forming cecids across phylogenetically diverse plant taxa. A potential bias in this observation should be noted, however, since many taxonomic works on cecids have assumed high host specificity. Future work may indicate that some species of cecids recorded from different single hosts may in fact turn out to be conspecific on a wide variety of hosts.

(3) The evolutionary steps to gall-forming appear complex. The gall-former has to "learn" to manipulate the defence reactions of the host plant, at the same time synchronizing its phenology with that of the plant. In our scenario, the evolution of the capacity to form galls evolved only once in the Cecidomyiidae, through an association with fungi that may already have possessed the gall-forming ability, with reversal to free-living fungal feeding

occurring a few times, and the phytophagous gall formers evolving from the ambrosia gall formers at least twice.

These conclusions bring to mind the discredited theory of the monophyletic origin of phytophagy and gall formation proposed by Möhn (1955, 1961). Möhn suggested that mycetophagy in the Cecidomyiinae evolved secondarily from phytophagy. For the reasons stated above, the gall-forming habit may be monophyletic, but derived from an endomycetophagous feeding mode.

ACKNOWLEDGEMENTS

We are grateful for the critical reviews of our manuscript provided by Drs J. Cumming, E. Lindquist, S. A. Redhead, and J. C. Roskam. Mr L. Forster mounted the specimens of Cecidomyiidae on microscope slides with his usual, masterful dexterity. We also thank Mr S. J. Darbyshire for assistance, particularly in sectioning of galls, photography, and plant identification.

REFERENCES

Anderson, N. M. (1982). "The Semiaquatic Bugs (Hemiptera, Gerromorpha). Phylogeny, adaptations, biogeography and classification". Scandinavian Science Press, Klampenborg, Denmark.

Baccarini, P. (1893). Sopra un curioso cecidio della *Capparis spinosa* L. *Malpighia* **7,** 405–414.

Batra, L. R. (1963). Ecology of ambrosia fungi and their dissemination by beetles. *Trans. Kans. Acad. Sci.* **66,** 213–236.

Batra, L. R. (1964). Insect-fungus blister galls of *Solidago* and *Aster*. *J. Kans ent. Soc.* **37,** 227–234.

Batra, L. R. and Lichtwardt, R. W. (1963). Association of fungi with some insect galls. *J. Kans. ent. Soc.* **36,** 262–278.

Borkent, A. and Bissett, J. (1985). Gall midges (Diptera: Cecidomyiidae) are vectors for their fungal symbionts. *Symbiosis* **1,** 185–194.

Bronner, R. (1977). Contribution a l'étude histochimique des tissus nourriciers des zoocécidies. *Marcellia* **40,** 1–134.

Buchner, P. (1930). "Tier und Pflanze in Symbiose". Borntraeger, Berlin.

Camp, R. R. (1981). Insect-fungus blister galls on *Solidago graminifolia* and *S. rugosa*. I. A macroscopic and light microscope study of the host-parasite relationship. *Can. J. Bot.* **59,** 2466–2477.

Docters van Leeuwen, W. M. (1929). Ueber eine Galle auf *Symplocos fasciculata* Zoll., verursacht durch eine Gallmücke: *Asphondylia bursaria* Felt, die mit einem Fungus zusammen lebt. *Marcellia* **25,** 61–66.

Docters van Leeuwen, W. M. (1939). An ambrosia-gall on *Symplocos fasciculata* Zoll. *Annu. Buitenzorg Lands Plantentium* **49,** 27–42.

Freeman, B. E. and Geoghagen, A. (1987). Size and fecundity in the Jamaican gall-midge *Asphondylia boerhaaviae*. *Ecol. Entom.* **12,** 239–249.

Gagné, R. J. (1968). A taxonomic revision of the genus *Asteromyia* (Diptera: Cecidomyiidae). *Misc. Publs ent. Soc. Am.* **6,** 1–40.

Gagné, R. J. (1973). Cecidomyiidae from Mexican Tertiary amber (Diptera). *Proc. ent. Soc. Wash.* **75,** 169–171.

Gagné, R. J. (1976). New Nearctic records and taxonomic changes in the Cecidomyiidae (Diptera). *Ann. ent. Soc. Am.* **69,** 26–28.

Gagné, R. J. (1977). Cecidomyiidae (Diptera) from Canadian amber. *Proc. ent. Soc. Wash.* **79,** 57–62.

Gagné, R. J. (1986). The transition from fungus-feeding to plant-feeding in Cecidomyiidae (Diptera). *Proc. ent. Soc. Wash.* **88,** 381–384.

Gagné, R. J. and Hawkins, B. A. (1983). Biosystematics of the Lasiopterini (Diptera: Cecidomyiidae: Cecidomyiinae) associated with *Atriplex* spp. (Chenopodiaceae) in southern California. *Ann. ent. Soc. Am.* **76,** 379–383.

Goidanich, A. (1941). Interpretazione simbiotica di una associazione micoentomatica gallare. *Atti Accad. Sci. Torino* **76,** 208–221.

Harris, K. M. and Evans, R. E. (1979). Gall development in the fungus *Peniophora cinerea* (Fr.) Cooke induced by *Brachyneurina peniophorae* sp. n. (Diptera: Cecidomyiidae). *Entomologist's Gaz.* **30,** 23–30.

Kaiser, P. (1978). Ueber die Bedeutung von Pilzhyphen in verpilzten Mückengallen, Untersuchungen an *Lasioptera rubi* (Schrank, 1803) (Diptera, Cecidomyiidae). *Ent. Mitt. zool. St. Inst. zool. Mus. Hamburg* **6**(99), 41–48.

Larew, H. G., Gagné, R. J., and Rossman, A. Y. (1987). Fungal gall caused by a new species of *Ledomyia* (Diptera: Cecidomyiidae) on *Xylaria enterogena* (Ascomycetes: Xylariaceae). *Ann. ent. Soc. Am.* **80,** 502–507.

Mamaev, B. M. (1964). Ecological aspects of the evolution of gall midges (Diptera, Itonididae-Cecidomyiidae). *Zool. Zh.* **43,** 1169–1180.

Mamaev, B. M. (1968). "Evolyutsiya Galloobrazuyushchikh Nasekomykh Gallits". Nauka, Leningrad.

Mamaev, B. M. (1975). "Evolution of Gall-Forming Insects—Gall Midges". (Trans. A. Crozy.) British Library, Boston Spa.

Mani, M. S. (1964). "Ecology of Plant Galls". W. Junk, The Hague.

Meyer, J. (1952). Cécidogenèse de la galle de *Lasioptera rubi* Heeger et rôle nourricier d'un mycélium symbiotique. *C. r. hebd. Séanc. Acad. Sci., Paris* **234,** 2556–2558.

Meyer, J. and Maresquelle, H. J. (1983). "Anatomie des Galles". Borntraeger, Berlin.

Möhn, E. (1955). Beitrage zur Systematik der Larven der Itonididae (=Cecidomyiidae, Diptera). I. Teil: Porricondylinae und Itonidinae Mitteleuropas. *Zoologica* **105** (1, 2), 1–247.

Möhn, E. (1961). Gallmücken (Diptera, Itonididae) aus El Salvador. 4. Zur Phylogenie der Asphondyliidi der neotropischen und holarktischen Region. *Senckenberg. biol.* **42,** 131–330.

Morse, J. C. and White, D. F. (1979). A technique for analysis of historical biogeography and other characters in comparative biology. *Syst. Zool.* **28,** 356–365.

Neger, F. W. (1909). Ambrosiapilze. II. Die Ambrosia der Holzbohrkafer. *Ber. dt. bot. Ges.* **27,** 372–389.

Neger, F. W. (1910). Ambrosiapilze III. weitere Beobachtungen an Ambrosiagallen. *Ber. dt. bot. Ges.* **28,** 455–480.

Orphanides, G. M. (1975). Biology of the carob midge complex, *Asphondylia* spp. (Diptera, Cecidomyiidae), in Cyprus. *Bull. ent. Res.* **65,** 381–390.

Price, P. W., Waring, G. L., and Fernandes, G. W. (1986). Hypotheses on the adaptive nature of galls. *Proc. ent. Soc. Wash.* **88,** 361–363.

Price, P. W., Fernandes, G. W., and Waring, G. L. (1987). Adaptive nature of insect galls. *Env. Ent.* **16,** 15–24.

Puzanowa-Malysheva, E. W. (1935). *Asphondylia prunorum* Wachtl (Diptera, Cecidomyiidae) un deren Pilzgallen am Pflaumenbaum. *Angew. Ent.* **21,** 443–462.

Roskam, J. C. (1985). Evolutionary patterns in gall midge-host plant associations (Diptera, Cecidomyiidae). *Tijdschr. Ent.* **128,** 193–213.

Ross, H. (1932). "Praktikum der Gallenkunde (Cecidologie)". Springer-Verlag, Berlin.

Schweinitz, L. D. von (1822). Synopsis fungorum Carolinae superioris. *Schr. naturf. Ges. Leipz.* **1,** 20–131.

Schweinitz, L. D. von (1832). Synopsis fungorum in American boreali media degentium. *Trans. Am. phil. Soc.* II, **4,** 141–316.

Skuhrava, M., Skuhravy, V., and Brewer, J. W. (1984). Biology of Gall Midges. In "Biology of Gall Insects" (T. N. Ananthakrishnan, ed.), pp. 169–222. Oxford and IBH Publishing, New Delhi.

Sylvén, E. (1979). Gall midges (Diptera, Cecidomyiidae) as plant taxonomists. *Symb. bot. upsal.* **22**(4), 62–69.

Trelease, W. (1884). Notes on the relations of two Cecidomyians to fungi. *Psyche Camb.* **4,** 195–200.

Trotter, A. (1900). Ricerche intorno agli entomocecidii della flora italica. *Nuovo G. bot. ital.*, n.s. **7,** 187–206.

Trotter, A. (1905). Nuove ricerche sui micromiceti delle galle e sulla natura dei loro rapporti ecologici. *Annls mycol.* **3,** 521–547.

Weis, A. E., Price, P. W., and Lynch, W. (1983). Selective pressures on clutch size in the gall maker *Asteromyia carbonifera. Ecology* **64**, 688–695.

Wood, D. M. and Borkent, A. (in press). Phylogeny and classification of the Nematocera. *In* "Manual of Nearctic Diptera" (J. F. McAlpine, D. M. Wood, N. E. Woodley and A. Borkent, eds), vol. 3. Agriculture Canada, Ottawa.

Yukawa, J., Ohtani, T., and Yazawa, Y. (1983). Host-change experiments from wild plants to soybean in *Asphondylia* species (Diptera, Cecidomyiidae). *Proc. Ass. Pl. Prot., Kyushu* **29,** 115–117.

however impressive the numerous hypotheses are (Howe and Smallwood, 1982), the study still suffers from incompleteness and imbalance not only in the fragmentary nature of published data (Janzen, 1983) but also in the failure to recognize fully that plants and animals are not the sole, or even principal, actors in this evolutionary drama (Smith, 1970; Herrera, 1985).

Most animals are merely food harvesters, processors, and fermentation tanks, "frills around a compost heap" (Janzen, 1985) because feeding is mediated by microbial "biochemical brokers" (Southwood, 1985). Evolution of feeding, as process and behaviour, is largely at the level of gut microbiota, which helps to explain why preferred food and food-processing organs are such conserved features in the history of metazoa. It is often overlooked that feeding in plants is also mediated by organisms of the rhizosphere: that plants "wear their guts on the outside" (Janzen, 1985).

"Guts" in most seed plants are mycorrhizal fungi. For obligately mycotrophic plants, a critical component of the nebulous "safe" site (Harper, 1977) for the establishment of reproducing progeny from seed must be the mycorrhizal fungus. A common generalization is that the establishment of safe sites for mycorrhizal plants is left to chance: a fortuitous placement, by biotic or abiotic agencies, of a seed near the fungus, or vice versa.

Mycotrophism in extant seed plants evolved mainly along two lines: vesicular-arbuscular mycorrhizae (VAM), and ectomycorrhizae (Malloch *et al.*, 1980). Present-day examples of stomach-motivated zoochory also followed two separate lines of evolution: where fruit is a separate resource offered as a reward for dispersal, and where some seeds of a seed crop are sacrificed to seed-eating animals as a price for the dispersal of others. In woody plants there appears to be a broad correlation between the respective pairs: those with VAM often bear palatable fruit, those with ectomycorrhizae bear palatable seed.

Here we explore the possibility that the correlation is real; that is, that the evolution of both seed-dispersal strategies was selected primarily by the advantages of uniting in a safe site two essential but biologically and topographically disparate—shoot-borne and root-borne—components. The problem of concomitant dispersal is as old as mycotrophism itself but was lessened when free-ranging animals were manipulated, via stomachs, into a mutualistic "ménage à trois". During the history of plant–animal interaction, the "compost heap" and the mycorrhizal "gut" remained fundamentally the same; the "frills" around them change in a relatively quick succession. The effective routine for establishing mycorrhizal symbioses is nearly the same, whether performed by ants, rodents, or man.

Most examples of endozoochory are viewed as two-way plant–animal mutualisms, and interpreted as being "diffuse" and mutually non-obligatory (Janzen, 1985). It is not obvious whether such relationships accrued from fortuitous congruence of independently evolved traits (coadaptation), or are the outcome of a sequence of reciprocal evolutionary adjustments in which the interacting organisms themselves act as selective agents (coevolution).

Our contention that seed dispersal by fruit- and seed-eating animals is a three-way interaction involving mycorrhizal fungi occasions re-appraisal of some current interpretations.

DISPERSAL OF VAM PLANTS

VAM, even when recognized as an integral component of plants, is rarely considered relevant to the three essentials of zoochory—who moves the seed, where seeds land, and what happens to landed seeds (Janzen, 1983)—let alone as motivating coevolution of plant and seed-dispersing animals.

The dispersal of VAM fungi too is usually treated as independent of the movement of seeds. VAM fungi spread locally by root contact, by movement of "infected" plant material, and by movement of spores by wind, water, or animals: invertebrates (MacMahon and Warner, 1984), as well as mammals, chiefly shrews, mice, voles, and pocket gophers (Fogel and Trappe, 1978; Ovaska and Herman, 1985). Spores dispersed by animals are usually the indigestible residue of a diet of VAM fungi, "infected" roots, or plant debris, or as in the case of shrews and mice, of mycophagous invertebrates (McIntire, 1984; Sieg *et al.*, 1986). The dependence of many VAM fungi for dispersal by vertebrate mycophagous animals shows in the evolution of the sclerocarp that, like fleshy fruit, is offered to animals as a reward for spore dispersal (Maser *et al.*, 1978). The same animals may be anatomically and behaviourally adapted for dispersal of plants in that they collect, move, and hoard below ground seeds or other organs of propagation.

Long-range dispersal of VAM fungi is attributed to birds, albeit as passive transit in "mud on feet" (McIlveen and Cole, 1976). However, migrating frugivorous birds are "passed from plant to plant all along the way" (Janzen, 1985). Plants offer conspicuously coloured fruit timed for seasonal migration (Howe and Smallwood, 1982; Izhaki and Safiel, 1985). Some plants are stripped of fruit in the fall, while the fruit of others is ignored (Herrera, 1985), presumably to await returning transients. Many such birds also consume invertebrates, which adaptation may be viewed as a diffusely coevolved mutualism imposed by discontinuous fruiting in individual plants and their inability to restrict the range of frugivores (Herrera, 1985; Janzen, 1985), or as a stage in the evolutionary transition towards obligate phytophagy (Southwood, 1985). A transient American robin hops from ground to trees, from pulling out an earthworm to picking a fruit, dispensing a concentrate of fruit-borne seed and earthworm-borne VAM spores.

However, although robins and other birds and mammals fulfill the requirements, concomitant dispersal may be an anachronism, in that transport of the fungus is no longer essential for the establishment of VAM plants in late successional and climax communities. Seeds of most VAM plants reliably find fungal partners and so are not now dependent on a "coevolutionary triangle" that would have controlled, with important input from birds (Regal, 1977), the early radiation of the angiosperms. If concomitant dispersal is still vigorously selected for, it would be in strategies aimed at colonization of virgin sites, which, today, result mainly from natural or man-made disturbances.

Colonization of disturbed sites

Pioneer colonizers of disturbed sites are usually non-mycorrhizal, *r*-selected fugitives (Miller, 1979) that are quickly succeeded (Allen and Allen, 1980; Medve, 1984) by more stable communities equipped to cope with a nutrient-impoverished environment by being mycotrophic (Reeves *et al.*, 1979).

The establishment of VAM plants on disturbances created by physical forces or by ground-breaking activities of animals, including man, on sites occupied by ectomycorrhizal plants, or where a severe environment leads to discontinuous vegetation, depends on pH, water, and nutrient status of the soil, and on the initial availability, and potential for subsequent introduction of VAM fungi (Doerr *et al.*, 1984; Trappe, 1986). In recent disturbances, barrens, and established stands of ectomycorrhizal plants, VAM fungi may be rare or absent (Kovacic *et al.*, 1984; MacMahon and Warner, 1984). Defaecation of viable spores of VAM fungi, as by grasshoppers, rabbits, or mice on mine spoils (Rothwell and Holt, 1978; Ponder, 1980) may be superior to physical transport in "activating" seed-bank seeds, or in fortuitously bringing together seeds and spores of VAM plants. However, the concomitant dispersal of seeds and spores by specific animals appears to be more significant in initiating a succession of VAM plants, even though at present this claim is supported only by anecdotal evidence.

Abandoned seed middens, especially of granivorous rodents, are sources of enhanced recruitment (Reichman, 1979) of VAM plants in a variety of disturbances (Batzli, 1975; Davidson *et al.*, 1984; McIntire, 1984; Dimitriev, 1985). Seed-hoarding granivores such as deer mice select seed for quality (Wolf and Dueser, 1986) and utility (Voight and Glenn-Lewin, 1979). Deer mice also defaecate spores of several different VAM fungi (Fogel and Trappe, 1978), which they presumably acquire from occasionally consumed roots, or, more likely, "second-hand" from soil animals: ground beetles, crickets and millipedes are among invertebrates that account for 50–75% of the deer mouse diet (Sieg *et al.*, 1986). Significantly, consumption of arthropods does not coincide with their greatest abundance but may supplement the diet of seeds to a greater extent than the diet of forbs. In addition, deer mice eat berries and defaecate viable seed (Wolf and Deuser, 1986), very likely together with viable spores of VAM fungi. Deer mice are ecologically adventurous, foraging in and between habitats as diverse as VAM and ectomycorrhizal forests, grasslands, and riparian communities (MacCracken *et al.*, 1985), often preferring marginal or disturbed sites (Wolf and Dueser, 1986; Voight and Glenn-Lewin, 1979). Other seed-hoarding granivorous/mycophagous mice, jumping mice, and voles may be as important in dispersing VAM plants, albeit within more restricted territories (Yahner, 1982). Also pikas, which colonize new disturbances, are credited with establishing VAM plants on such sites in montane North America and arctic and semi-arid Eurasia (Dimitriev, 1985). An indirect but effective concomitant dispersal of seeds and spores may be performed by raptorial birds (Balgooyen and Moe, 1973) and by mammalian carnivores.

Judging from the multitude of opportunities and adaptations for concomitant dispersal of VAM plants, and from their competitive superiority, the manifestly transient, *r*-selected, non-mycorrhizal species (Reeves *et al.*, 1979) or ecotypes of facultative mycotrophs (Miller, 1979) could be at an evolutionary disadvantage compared with mycorrhizal plants. However, this is inconsistent with claims of the relatively recent origin of ruderals and advanced autonomy of such "general-purpose genotypes" (Baker, 1970), and is negated by the existence of stable populations (Platt, 1975, 1976; Davidson and Morton, 1981) or communities (Buckley, 1982) of non-mycorrhizal plants. It is therefore pertinent to examine the possibility that adaptations of plants, fungi, and animals towards concomitant dispersal of VAM plants are counteracted by adaptations aimed at prolonging dominance of non-mycorrhizal plants through maintenance of animal-generated disturbance, of which man is the chief perpetrator.

NON-MYCORRHIZAL PLANTS AND CHRONIC DISTURBANCE

Soil excavated by burrowing animals is colonized by weedy non-mycorrhizal plants that may be conspicuously different from those on adjacent undisturbed soil, and that persist as long as the burrows are inhabited. In arctic and alpine habitats, lemmings, ground squirrels, and marmots are chiefly responsible for such "vegetational mosaics" (Tikhomirov, 1959; Del Moral, 1984). In arid lands the activities of excavating animals such as ground squirrels, voles (Naumov, 1975), badgers (Platt, 1976), and ants may lead to more stable communities of non-mycorrhizal plants. This is especially true of

Australian heaths, where ants are the sole agents of continuous disturbance (Davidson and Morton, 1981; Buckley, 1982). Excavated soil may be inimical to establishment of mycorrhizal symbiosis because of an inadequate complement of such fungi and its physical and chemical properties (e.g., "salt earth", Naumov, 1975; low moisture, Lamont, 1982), including a relatively high initial fertility. Quick-flowering non-mycorrhizal species or ecotypes would therefore be at an advantage in escaping competition, even from mycorrhizal conspecifics. But, although a rapid, opportunistic filling of the niche by non-mycorrhizal plants may slow the start of a secondary succession of mycorrhizal ones (Allen and Allen, 1980), by preventing multiplication of fortuitously dispersed mycorrhizal fungi, when burrows are abandoned the vegetation reverts to that characteristic of the undisturbed sites, presumably because of rapid impoverishment of soil.

The maintenance of highly diverse disturbance-borne non-mycorrhizal floras, at least in mesic regimes, consequently appears to depend primarily on continuous enrichment by excrement and "house cleaning" debris of soil-excavating animals, whether mammals (Tikhomirov, 1959; Batzli, 1975) or ants (Buckley, 1982; Beattie, 1985). "Herbivore control of nutrient flux" (Batzli, 1975) is a special type of "direct nutrient cycling" (Went and Stark, 1968) in which plants replaced mycorrhizal fungi with sedentary, social animals. It is not, therefore, surprising that plants compete for such "safe" sites, through the stomachs of animals that maintain them (Dinesman, 1967; Heithaus, 1981; Davidson *et al.*, 1984) and that some interrelationships are coevolved, even if the coevolutionary response of the animals appears to be primarily behavioural.

Successful zoochory is to get "the seeds into the right animals and to keep them out of the wrong animals" (Janzen, 1983), whether as a fleshy fruit or a share of the seed crop. That burrow-associated plants evolved different seed and seeding characteristics, attracting more specific mammalian dispersers, can be inferred from the evidence of habitat and resource partitioning among granivorous rodents (Bowers, 1982; Kelrick *et al.*, 1986). Specific adaptations of myrmecochores include seed size, display, and seeding phenology and, more characteristically, an edible elaiosome coupled to an inedible seed (Buckley, 1982). The mere presence of plant adaptations enticing dispersers, however specialized, does not entail coevolution. However, it is difficult to explain in terms of fortuitous coadaptations the "weeding out" by voles of woody plants growing in the vicinity of burrows or otherwise impinging on "cultivated" disturbances (Naumov, 1975; Bergman, 1982), or a similar selective weeding performed by ants, or removal by ants of elaiosomes only after the inedible attached seeds have been carried to the nest, only to be discarded, sometimes into protected middens on or around the nest (Buckley, 1982).

The patterns emphasized above are broad correlations to be tested by individual case studies. Not all animals that create and maintain disturbance select for non-mycorrhizal plants, nor do all seed-hoarding granivores that also ingest VAM fungi disperse VAM plants. Harvester ants that remove germinated seeds from middens and "plant" them around the nest (Trevis, 1958), or male rats that store seeds and bulbs in underground chambers (Galil, 1967) may promote the establishment of mycorrhizal plants and prevent establishment of non-mycorrhizal ones. The picture is complicated further if the ecologically versatile facultatively mycotrophic plants are considered.

FACULTATIVE MYCOTROPHS, GRAZERS, AND GRANIVORES

Recent debate on the origin and nature of the interdependence of grasses and grazers (Belsky, 1986) has confirmed that the association is coevolved (e.g., silica versus teeth), but is divided as to whether the interaction is antagonistic or mutualistic, especially since McNaughton (1984) demonstrated that heavily grazed pastures gain in biomass and diversity. However, coevolved interdependence of grasses and grazers is more likely to be evident in adaptations for seed dispersal. Janzen (1983, 1984) claimed that the usual and selected mode of dispersal in grasses and forbs is through consumption of seeds together with vegetative parts that serve as "fruit". Undoubtedly, some forbs and grasses are coprophilous in that, like cyclic coprophilous fungi, they originate from defaecated diaspores, grow on dung or its former site, and set seed adapted for ingestion and passage through large herbivores to complete the cycle. A species of *Sporobolus* that grows out as a pure stand from incubated dung of cattle grazed on Canadian prairies may be an example (personal observation). The consumption of small, indigestible seed could be a coevolved dispersal mutualism when the vegetative portion of the plant is the staple diet of the animal and when that portion is dispensable or redundant. This interpretation, however, poses additional questions; for example, what are the implications of the deposition of seeds in a large parcel of fertilizer for a mycotrophic plant? The circumstances would probably select for facultatively mycorrhizal plants in which transient r-strategy is succeeded by K-strategy when dung nutrients have been dissipated. An example of this may be seen in wild species of strawberries that can often be found germinating in dung and can grow into the soil and become established (personal observation). Malloch and Malloch (1982) suggested that the Rosoideae, to which strawberries belong, may be facultatively mycotrophic. The transition from antagonistic seed predation to mutualistic dispersal by

tain ectotrophs have coevolved with certain rodents and corvids is indicated by reciprocal anatomical and behavioural adaptations, emphasized by Smith (1970), Lanner (1982) and Smith and Reichman (1984). The evolutionary paths of tamiascurine squirrels, jays, and nutcrackers, and those of the Cembrae, Strobi, and Cembroides pines particularly appear to have been closely intertwined (Tomback, 1983), at least since the Miocene (Smith, 1970).

Dispersal of spores

The squirrels, voles, and mice that eat, cache, and disperse seeds of northern ectomycorrhizal trees also eat ectomycorrhizal fungi and defaecate viable spores (Fogel and Trappe, 1978; Gronwall and Pehrson, 1984; Hunt and Maser, 1984), often in astronomical numbers (Kotter and Farentinos, 1984).

Documented cases of avian mycophagy are more rare, but jays can eat macrofungi that they sometimes steal from caches of squirrels (Alsheikh and Trappe, 1983).

The relationship between mycophagous rodents and hypogeous fungi also suggests a coevolved mutualism: in exchange for dispersal of spores, these fungi offer ripe "fruit" advertised by scent. The attractant appears to be a mammalian sex pheromone (Claus *et al.*, 1981), a potent incentive for arboreal seed eaters to "come down to earth" where they would be most vulnerable to predators. The evolution of seed burial may originate in the advantages to an animal of caching in soil where antibiotic action of the rhizosphere could suppress microbial degradation of seeds (Smith and Reichman, 1984). Placement of seeds near antibiotic producing ectomycorrhizal sheaths could be particularly advantageous, and lead to the evolution of traits facilitating dispersal of ectotrophs to "safe" sites.

What makes a "safe" site safe?

Harper's (1977, p. 466) statement that "it may be that it is the act of seed caching that determines which seed among millions is the one that produces the descendant tree" is supported in subsequent literature (Smith and Reichman, 1984; Jensen, 1985). Bossema (1979) found that acorns he planted in jays' territory resulted in seedlings that were invariably less well rooted than those sprouting from jays' caches. The seedlings "planted" by jays were very probably mycorrhizal, but neither Bossema nor his fellow ecologists discussed mycotrophism. On the other hand, mycologists view mycophagy and the habit of burrowing rodents defaecating over and among fine roots as adaptations promoting the establishment of ectomycorrhizal symbiosis and, should the animals dwell at the edge of a forest, of enhancing colonization of new areas if droppings were to be fortuitously placed near seeds or vice versa (Maser *et al.*, 1978). Such independent endozootic dispersal of spores has no particular advantage over concomitant dispersal of seeds and spores by wind, the strategy adopted by spruces and the Betulaceae. The maintenance by other ectotrophs of energetically expensive coteries of seed-eating vertebrate dispersers must have another advantage, which prompts us to consider the possibility that mycophagy and seed burial are but two sides of the same evolutionary coin.

It is not known (a) whether animals place seeds or spores near each other or disperse them together; (b) whether spores are defaecated as an accompaniment to voiding of transported seeds or to mark caches; (c) whether hypogeous fungi and seeds are harvested, transported and cached together; or (d) whether seeds are placed in holes from which hypogeous fungi have been excavated. Unlike avian cheaters whose diet is often confined to seeds, jays and nutcrackers eat invertebrates and, perhaps, also acquire fungal spores via mycophagous beetles, their larvae, or slugs, which themselves serve as local vectors of mycorrhizal inoculum (Richter, 1980; Crowson, 1984).

However, it is known that how, where, and what seeds are cached facilitates the establishment of ectomycorrhizal symbiosis.

1. Seeds buried in caches are those of ectomycorrhizal trees (Sork, 1983; Tomback, 1983; Nilsson, 1985), whereas the same animals may hoard supplementary food items derived from VAM plants arboreally (Kato, 1985). Unlike avian and mammalian cheaters, jays and nutcrackers avoid insect-infested seeds even though they supplement seed diet with invertebrates (Smith and Balda, 1979). Only the fittest acorns, nuts (Bossema, 1979; Sork, 1983), or pine cones and seeds (Vander Wall and Balda, 1977; Lanner, 1982; Tomback, 1983) are selected for caching. Rodents and birds perform a further selection by visiting sprouting caches (to feed on cotyledons, for example), and pulling out all but the sturdiest, presumably mycorrhizal, seedlings (Abbott and Quink, 1970; Bossema, 1979).
2. For many vertebrate-dispersed ectotrophs, burial of seed is often necessary for, or enhances, germination (Ovington and MacRae, 1960; Jensen, 1985).
3. Seeds are normally cached within ectomycorrhizal forests (Tomback, 1983) or their immediate vicinity (Bossema, 1979), that is, not only where seeds best escape robbers and cheaters (Smith and Reichman, 1984), but also where seedlings are most likely to survive (Mattes, 1982) by encountering mycorrhizal fungi.

However, even the concomitant dispersal of seeds and spores by specialized animals within ectotroph-dominated habitats cannot ensure establishment. Ectomycorrhizal plants associate with different fungi at different spatial and temporal stages of succession (Dighton and Mason, 1985). Seeds that are routinely cached under parent trees or established conspecifics (Vander Wall and Balda, 1977) have little chance of establishment, not because of predation but because of seedling failure (Bossema, 1979; Nilsson, 1985). In contrast, seeds cached on sites from which trees have been removed by man or natural catastrophes are the chief source of "natural" replanting (Ligon, 1978).

It is debatable whether seedling recruitment limited to filling of the existing (West, 1968) or newly created (Chim and On, 1973) gaps in established forests or their peripheries (Bossema, 1979) is so much more effectively mediated by seed-caching vertebrates than by invertebrates, or even by wind, to justify the cost. The cost in seeds for the establishment of occasional ectomycorrhizal "replacements" by vertebrates is colossal (Abbott and Quink, 1970), without counting that in most years the entire seed output is needed to maintain a coterie of seed-caching dispersers (Ligon, 1978; Sork, 1983; Jensen, 1985). Admittedly "only one successful seedling on average is expected to establish from a whole lifetime of seed production by the parent plant" (Harper, 1977, p. 465). But why do most ectomycorrhizal plants inflate the cost by shedding energy-rich edible seeds, accessible to all? Clearly, a fleshy fruit containing inedible seed has been strongly selected against in ectotrophs generally, possibly because unilateral endozootic dispersal by frugivores does not meet the requirements of a geographically more discrete and ecologically specialized ectomycorrhizal symbiosis. Toxic seeds evolved in tropical ectomycorrhizal legumes, albeit at the price of the plant's mobility. The same may be true of those of the northern ectotrophs (e.g., spruce) that opted for small, more toxic, abiotically dispersed seed, and are geographically less adventurous or successionally later than their corvid- and rodent-dispersed relatives (Howe and Smallwood, 1982; Jensen, 1985; Nilsson, 1985).

Like nectar and pollen that are offered as a reward for pollination, the edible seed as the currency of "payment in offspring" seed-dispersal strategy (Janzen, 1985) cannot be viewed as accidental but rather signals coevolved mutualism. The advantage that, for ectotrophs, commands such a high price in offspring is distance dispersal; it is realized through mast-fruiting!

Mast-fruiting

Supra-annual fruiting synchronized at community or population level is characteristic of virtually all extratropical ectotrophs as well as tropical Dipterocarpaceae (Janzen, 1974, 1978). In any one area, fruiting may be synchronized between communities of conspecifics or congenerics (Smith, 1970) but members of different genera usually mast at different times (Barnett, 1977; Jensen, 1985) in response to as yet unidentified cues. The common hypothesis of an environmental cue does not convincingly account for *Pinus cembra*, for example, whose community-wide mast may cover thousands of square kilometers. Perhaps such vast but continuous expanses communicate by pheromones (Janzen, 1978) or through the network of ectomycorrhizal fungi.

Mast fruiting is considered to be a result of coevolution with animals and is presumed to have started in response to a freak environmental event, and to be fine-tuned by predators that eliminate seeds of individuals that fruit out-of-phase. Consequently, mast-fruiting is almost universally interpreted as a strategy for escaping seed predation "in time" through sporadic community-wide satiation of predators (Janzen, 1971; Howe and Smallwood, 1982). Indeed, mast satiates seed predators, especially insects adapted for taking seeds *in situ* prior to dispersal (Silvertown, 1980), but there is an important distinction between these and seed predators that also act as dispersers and especially burial agents as well (Harper, 1977, p. 457; Smith and Balda, 1979). The latter create for plants a conflict, in that reduced population of seed predators resulting from mast is at the expense of seed dispersal (Stepanian and Smith, 1978).

We interpret the combination of seed-caching seminivores and mast-fruiting ectotrophs not as a conflict of strategies but as a coevolved compromise. As seed predators, these animals may contribute to the sharpening of periodicity of mast-fruiting, but that is probably incidental. Wind-dispersed southern beeches fruit supra-annually in synchrony and the northern ectotrophs of New Guinea mast-fruit without the benefit of nut- and large seed-eating vertebrates. To deter vertebrate predators, seeds need to be indigestible, toxic, or otherwise unpalatable. If seeds are edible, they are more likely adapted to animals whose feeding behaviour is compatible with seeding requirements of the plants. We therefore suggest that edible seeds and their dispersal by seed-eating vertebrates are coevolved traits that co-adapted with mast-fruiting to result in a more effective dispersal by the following methods.

1. *Increasing the population size of dispersers.* Supranumerary out-of-season "double breeding" of pinyon jays (Ligon, 1978) or bearded pigs (Caldecott and Caldecott, 1985) is not in response to the bounty of mast but in anticipation of it! Breeding is triggered by superabundance of green cones or flowers respectively, to yield juvenile dispersers (which,

as a bonus, destroy fewer seeds; Smith and Reichman, 1984) in time for the bumper crop of seeds.

2. *Attracting immigrants and their hoarding*. Mast-fruiting populations or communities of conspecific or more distantly related ectotrophs attract animals, dispersers and predators, by causing relaxation of normal feeding behaviour, food preferences, habitat specialization, and bonds of territoriality (Smith and Balda, 1979; Caldecott and Caldecott, 1985; Nilsson, 1985). One outcome of such "outcropping" may be genotypic diversity of seed caches (as in "multitrunk" pines: Tomback, 1983); also, the planting of seeds where their presence reverses normal successional trends (Harper, 1977, p. 465), which is baffling perhaps only when viewed from the perspective of a human life-span.
3. *Increasing the amount of seed dispersed by residents*. Mast-fruiting results in a large surplus of germinable seeds stored in soil, as nutcrackers (Lanner, 1982), jays, and rodents (Ligon, 1978) cache 3–5 times more seed than they need. The significant outcome is "satiation of caches" (Jensen, 1985) and, for birds, "satiation" of spatial memory.
4. *Increasing the distance of dispersal by residents*. During mast-years, mammals that scatter-hoard seeds at constant density and birds expand cache territories, sometimes into unusual, non-forest habitat (Bossema, 1979; Darley-Hill and Johnson, 1981; Nilsson, 1985).

Mast, pigs and concomitant dispersal

The "natural" consumers of fagaceous and dipterocarp mast in the Old World are pigs (Janzen, 1971, 1974; Caldecott and Caldecott, 1985). This association probably represents a coevolved seed-dispersal mutualism (Janzen, 1985), but the enormous cost in seeds (Genard and Lescourret, 1985) is more likely to be the price of concomitant dispersal with mycorrhizal fungi. Pigs also eat roots or, more likely, the associated fungi for which their proverbial penchant is now justified—truffles exude boar's sex pheromones (Claus *et al.*, 1981)! In northern ectomycorrhizal forests, the heaviest production of fungal sporocarps in the autumn coincides with higher mass of mycorrhizal roots (Vogt *et al.*, 1981) and, in turn, with fruiting generally and with mast-fruiting in particular. Therefore, it may well be that it is the mycorrhizal fungi that point pigs in the right direction, drive them "to exhaustion" (Caldecott and Caldecott, 1985), or even induce puberty in females by mimicking the "boar effect" (Booth, 1987).

There is little justification for interpreting winged seeds of dipterocarps and some pines as adaptations for wind dispersal when such seeds, like acorns and nuts, simply drop to the ground below; even less, when their food value is on par with that of seeds offered to animals as reward for dispersal. More likely, the shedding of edible, nutritious seeds was originally intended for an earth-bound seed predator–disperser. Indeed, for at least the Fagaceae and Dipterocarpaceae, the coevolved dispersers were probably pigs and peccaries until their widespread decline. The demise of the dipterocarp-associated pigs, especially in continental Asia, may have been relatively recent and influenced by man (Janzen, 1974), and left Malayan dipterocarps in a state of static "vegetation" because neither the native squirrels nor the birds disperse seed (MacKinnon, 1978). Mast-fruiting appears to remain the principal method of seedling recruitment (Chim and On, 1973). On the other hand, dispersal of Fagaceae may have been increasingly performed by squirrels and birds since the Miocene, except in the Southern Hemisphere where *Nothofagus* may be a fagalian archetype antedating coevolution with seed-dispersing vertebrates; it has tiny winged, apparently wind-dispersed seeds that cannot, however, establish progeny more than 40 metres beyond parent stands (Baylis, 1980).

However, there is no evidence that rodents or birds routinely disperse seeds and spores together, even if Clark's nutcrackers have been known to establish pinyon pines on isolated rocky ledges or in pumice up to 22 kilometres from the source of joint inoculum (Tomback, 1983). That the establishment of disjunct populations of ectotrophs requires concomitant dispersal of seeds and spores is a justified conclusion (Baylis, 1980; Malloch *et al.*, 1980). That no strategy for long-distance concomitant dispersal has yet succeeded is witnessed by the ectotrophs' territorial cohesiveness despite millions of years of experimentation (Pirozynski, 1983).

Overcoming constraints imposed on dispersal by mortality of seedlings within established communities and failure of seeds (owing to absence of suitable fungi) outside them, calls for drastic measures to achieve even stepwise expansion of a population. Synchronous mast-fruiting in itself is one such strategy, the origin of which may be more closely linked with mycorrhizal fungi than with the associated herbivores. Another may be "group suicide" by fire in some pines (Smith, 1970) and eucalypts (O'Dowd and Gill, 1984). However, there can be little doubt that the most effective strategy involves both mast-fruiting and seed-eating vertebrate dispersers, even if annual territorial gains for plants are metres rather than kilometres (Smith and Reichman, 1984; Jensen, 1985; Nilsson, 1985). For ectotrophs especially, the movement of seed is not (as Janzen (1985) claimed) a by-product of zoochory but its objective. As movement is the function of dispersers' autonomy, the "diffuse seed-dispersal mutualism" may be the best coevolutionary compromise, however wasteful the interaction may be. Indeed if "diffuse" mutualisms are characteristic of zoochory generally, it may be because, unlike pollination, seed dispersal does not involve the seed alone.

Janzen, D. H. (1984). Dispersal of small seeds by big herbivores: foliage is the fruit. *Am. Nat.* **123,** 338–353.

Janzen, D. H. (1985). The natural history of mutualisms, In "The Biology of Mutualism" (D. H. Boucher, ed), pp. 40–99. Croom Helm, London and Sydney.

Jensen, T. S. (1985). Seed–seed predator interactions of European beech, *Fagus silvatica* and seed eating animals. *Oikos* **44,** 149–156.

Kamil, A. C. and Balda, R. P. (1985). Cache recovery and spatial memory in Clark's nutcrackers (*Nucifraga columbiana*). *J. exp. Psychol. Anim. Behav. Proc.* **11,** 95–111.

Kato, J. (1985). Food and hoarding behavior of Japanese squirrels. *Jap. J. Ecol.* **35,** 13–20.

Kelrick, M. I., MacMahon, J. A., Parmenter, R. R., and Sisson, D. V. (1986). Native seed preferences of shrub-steppe rodents, birds and ants: the relationships of seed attributes and seed use. *Oecologia* **68,** 327–337.

Kotter, M. M. and Farentinos, R. C. (1984). Formation of ponderosa pine ectomycorrhiza after inoculation with feces of tassel-eared squirrels. *Mycologia* **76,** 758–760.

Kovacic, D. A., St. John, T. V., and Dyer, M. I. (1984). Lack of vesicular-arbuscular mycorrhizal inoculum in a ponderosa pine forest. *Ecology* **65,** 1755–1759.

Lamont, B. (1982). Mechanisms for enhancing nutrient uptake in plants, with particular reference to mediterranean South Africa and Western Australia. *Bot. Rev.* **48,** 597–689.

Lanner, R. M. (1982). Adaptation of whitebark pine for seed dispersal by Clark's nutcracker. *Can. J. For. Res.* **12,** 391–402.

Ligon, J. D. (1978). Reproductive interdependence of pinon jays and pinon pines. *Ecol. Monogr.* **48,** 111–126.

MacCracken, J. G., Uresk, D. W., and Hansen, R. M. (1985). Rodent–vegetation relationships in Southern Montana. *Northwest Sci.* **59,** 272–278.

MacKinnon, K. S. (1978). Stratification and feeding differences among Malayan squirrels. *Malay. Nat. J.* **30,** 593–608.

MacMahon, J. A. and Warner, N. (1984). Dispersal of mycorrhizal fungi: processes and agents. In "VA Mycorrhizae and Reclamation of Arid and Semi-Arid Lands" (S. E. Williams and M. F. Allen, eds), pp. 28–41. Wyoming University, Laramie.

Malloch, D. and Malloch, B. (1982). The mycorrhizal status of boreal plants: additional species from northeastern Ontario. *Can. J. Bot.* **60,** 1035–1040.

Malloch, D. W., Pirozynski, K. A., and Raven, P. H. (1980). Ecological and evolutionary significance of mycorrhizal symbioses in vascular plants. (A review). *Proc. natn. Acad. Sci. U.S.A.* **77,** 2113–2118.

Maser, C., Trappe, J. M., and Nussbaum, R. A. (1978). Fungal–small mammal interrelationship with emphasis on Oregon coniferous forests. *Ecology* **59,** 799–809.

Mattes, H. (1982). Die Lebensgemeinschaft von Tannenhäler und Arve. *Ber. Eidg. Anst. forstl. Versuchs.* **241,** 3–74.

McIlveen, W. D. and Cole, H. (1976). Spore dispersal of Endogonaceae by worms, ants, wasps and birds. *Can. J. Bot.* **54,** 1486–1489.

McIntire, P. W. (1984). Fungus consumption by the Siskiyou chipmunk within a variously treated forest. *Ecology* **65,** 137–146.

McNaughton, S. J. (1984). Grazing lawns: animals in herds, plant form and coevolution. *Am. Nat.* **124,** 863–886.

Medve, R. J. (1984). The mycorrhizae of pioneer species in disturbed ecosystems in Western Pennsylvania. *Am. J. Bot.* **71,** 787–794.

Miller, R. M. (1979). Some occurrences of VA mycorrhizae in natural and disturbed ecosystems of the Red Desert. *Can. J. Bot.* **57,** 619–623.

Naumov, N. P. (1975). The role of rodents in ecosystems of the northern deserts of Eurasia. In

"Small Mammals: Their Productivity and Population Dynamics" (F. B. Golley, K. Petrusewicz and L. Ryszkowski, eds), pp. 299–309. Cambridge University Press, Cambridge.

Nilsson, S. G. (1985). Ecological and evolutionary interactions between reproduction of beech *Fagus silvatica* and seed eating animals. *Oikos* **44,** 157–164.

O'Dowd, D. J. and Gill, A. M. (1984). Predator satiation and site alteration following fire: mast reproduction of alpine ash (*Eucalyptus delegatensis*) in southeastern Australia. *Ecology* **65,** 1052–1066.

Ovaska, K. and Herman, T. B. (1985). Fungal consumption by six species of small mammals in Nova Scotia. *J. Mammal.* **67,** 208–211.

Ovington, J. D. and MacRae, C. (1960). The growth of seedlings of *Quercus petraea*. *J. Ecol.* **48,** 549–555.

Pirozynski, K. A. (1983). Pacific mycogeography: an appraisal. *Aust. J. Bot., Suppl.* **10,** 137–159.

Pirozynski, K. A., Carter, A., and Day, R. G. (1984). Fungal remains in Pleistocene ground squirrel dung from Yukon Territory, Canada. *Quarternary Res.* **22,** 375–382.

Platt, W. J. (1975). The colonization and formation of equilibrium plant species associations on badger disturbances in tall-grass prairie. *Ecol. Monogr.* **45,** 288–305.

Platt, W. J. (1976). The natural history of a fugitive prairie plant (*Mirabilis hirsuta* (Pursh.) MacM.). *Oecologia* **22,** 399–409.

Ponder, F. (1980). "Rabbits and Grasshoppers: Vectors of Endomycorrhizal Fungi on New Coal Mine Spoil. [Research Note no. NC250.] United States Department of Agriculture Forest Service.

Reeves, F. B., Wagner, D., Moorman, T., and Kiel, J. (1979). The role of endomycorrhizae in revegetation practices in the semiarid West. I. Comparison of incidence of mycorrhizae in severely disturbed vs. natural environments. *Am. J. Bot.* **66,** 6–13.

Regal, P. J. (1977). Ecology and evolution of flowering plant dominance. *Science, N.Y.* **196,** 622–629.

Reichman, O. J. (1979). Desert granivore foraging and its impact on seed densities and distribution. *Ecology* **60,** 1085–1092.

Richter, K. O. (1980). Evolutionary aspects of mycophagy in *Ariolimax columbianus* and other slugs. *Proc. Int. Colloq. Soil Zool.,* **7,** 611–636.

Rothwell, F. M. and Holt, C. (1978). "Vesicular-arbuscular mycorrhizae established with *Glomus fasciculatus* spores isolated from the faeces of cricetine mice." [Research Note no. NE 259.] United States Department of Agriculture Forest Service.

Sieg, C. H., Uresk, D. W., and Hansen, R. M. (1986). Seasonal diets of deer mice on bentonite mine spoils and sagebush grasslands in southern Montana. *Northwest Sci.* **60,** 81–89.

Silvertown, J. W. (1980). The evolutionary ecology of mast seeding in trees. *Biol. J. Linn. Soc.* **14,** 235–250.

Smith, C. C. (1970). The coevolution of pine squirrels (*Tamiasciurus*) and conifers. *Ecol. Monogr.* **40,** 349–371.

Smith, C. C. and Balda, R. P. (1979). Competition among insects, birds and mammals for conifer seeds. *Am. Zool.* **19,** 1065–1083.

Smith, C. C. and Reichman, O. J. (1984). The evolution of food caching by birds and mammals. *A. Rev. ecol. Syst.* **15,** 329–351.

Snow, D. W. (1981). Co-evolution of birds and plants. In "The Evolving Biosphere" (P. L. Forey, ed), pp. 169–178, Cambridge University Press, Cambridge.

Sork, V. L. (1983). Mammalian seed dispersal of pignut hickory during three fruiting seasons. *Ecology* **64,** 1049–1056.

Southwood, T. R. E. (1985). Interactions of plants and animals: patterns and processes. *Oikos* **44,** 5–11.

Stepanian, M. A. and Smith, C. C. (1978). A model for seed scatter-hoarding: coevolution of fox squirrels and black walnuts. *Ecology* **59,** 884–896.

Tikhomirov, B. A. (1959). "Relationships of the Animal World and the Plant Cover of the Tundra". [Transl. E. Issakoff and T. W. Berry.] Botanical Institute, Academy of Sciences of the USSR, Moscow and Leningrad.

Tomback, D. F. (1983). Nutcrackers and pines: coevolution or coadaptation? In "Coevolution" (M. H. Nitecki, ed.), pp. 179–223. University of Chicago Press, Chicago and London.

Trappe, J. M. (1986). Mycorrhizal fungi as determinants of plant successional patterns. *Newsl. mycol. Soc. Am.* **37,** 45.

Trevis, L. (1958). Interactions between the harvester ant *Vermessor pergandei* (Mayr) and some desert ephemerals. *Ecology* **39,** 695–704.

Vander Wall, S. B. and Balda, R. P. (1977). Coadaptations of the Clark's nutcracker and the piñon pine for efficient seed harvest and dispersal. *Ecol. Monogr.* **47,** 89–111.

Vinson, S. B. (1972). Imported fire ant feeding on *Paspalum* seeds. *Ann. ent. Soc. Am.* **65,** 988.

Vogt, K. A., Edmonds, R. L., and Grier, C. C. (1981). Biomass and nutrient concentration of sporocarps produced by mycorrhizal and decomposer fungi in *Abies amabilis* stands. *Oecologia* **50,** 170–175.

Voight, J. and Glenn-Lewin, D. C. (1979). Strip mining, *Peromyscus* and other small mammals in southern Iowa. *Proc. Iowa Acad. Sci.* **86,** 133–136.

Went, F. W. and Stark, N. (1968). The biological and mechanical role of soil fungi. *Proc. natn. Acad. Sci. U.S.A.* **60,** 497–504.

West, N. E. (1968). Rodent-induced establishment of ponderosa pine and bitterbush seedlings in central Oregon. *Ecology* **49,** 1009–1011.

Wolf, J. O. and Dueser, R. D. (1986). Noncompetitive coexistence between *Peromyscus* species and *Clethrionomys gapperi*. *Can. Fld Nat.* **100,** 186–191.

Yahner, R. H. (1982). Microhabitat use by small mammals in farmstead shelterbelts. *J. Mammal.* **63,** 440–445.

11 Coevolution by Horizontal Gene Transfer: A speculation on the role of fungi

K. A. PIROZYNSKI

Paleobiology Division, National Museum of Natural Sciences, Ottawa, Canada

Abstract

Multicellular organisms owe their complexity to intergenomic as well as intragenomic exchanges. Novel traits, when environmentally induced, may result from the rearrangement of transposable genetic elements; when induced by symbionts, they may result from the acquisition of "alien" genes. Microorganisms, including fungi, are potentially a direct source of plant genes, and may serve as gene vectors between interacting plants and animals. Gene transfer, such as characterizes agrobacterial tumorigenesis in plants may well accompany induction of plant galls by animals and fungi. Genetic canalization of symbiont-induced neoplasms could therefore lead to quantum evolutionary innovations, of which flowers, fruits, and organs of food storage and vegetative propagation are possible examples. Extrapolating from data pertaining to the integration of endophytic fungi, it is not inconceivable that the transition from a pathological neoplasm to a heritable neoseme could have involved reciprocal sequential suppression of sexual reproduction in both symbionts, possibly via mitochondrial plasmids.

INTRODUCTION

In contributions to this volume, as in most writings on coevolution, coevolution is viewed as a stepwise adjustment of interacting evolutionary units to reciprocal genetic change (Van Valen, 1983), each step entailing a "vertical" (Rose and Doolittle, 1983), usually sexual, transmission to progeny of novel inheritable traits.

Here, coevolution is explored from a more Lamarckian perspective: the ac-

COEVOLUTION OF FUNGI
WITH PLANTS AND ANIMALS
ISBN 0-12-557365-0

quisition of inheritable characteristics via a non-sexual, "horizontal" transfer of genes between distantly related, "reproductively isolated" organisms.

The evidence for much of what follows is at best circumstantial; its interpretation largely guided by the more speculative segments of pertinent literature, but encouraged by recent insights in molecular biology. Molecular biology has yet to provide the missing empirical data; nevertheless it has revealed situations that have surpassed even the wildest speculations. In presenting the following discussion I hope to stimulate further molecular investigations that could prove to throw new light on evolution within vascular plants.

GENETIC OPPORTUNITIES IN SYMBIOSIS

"... organisms are not limited for their evolution to genes that belong to the gene pool of their species ... Organisms and genomes may thus be regarded as components of the biosphere through which genes in general circulate at various rates and in which individual genes and operons may be incorporated if of sufficient advantage."

Jeon and Danielli (1971)

The existing intracellular symbioses involving unicellular hosts show degrees of biological integration that encompass "all those processes whereby two organisms of divergent evolutionary origins come together and exist as a stable, self-regulating unit" (Trench, 1980). The extent of integration appears to reflect the functional climax in a mutualistic equilibrium (e.g., photosynthetic inhabitants usually retain a greater degree of autonomy than non-photosynthetic ones; Taylor, 1983), as well as a temporal stage in a progressive loss of redundancies leading to the irreversible loss of genetic information from the inhabitant (Margulis, 1980; Taylor, 1983). Many associations have progressed to a point at which the inhabitant is essential for the survival of the exhabitant and has become, to all intents and purposes, the exhabitant's organelle (Jeon and Jeon, 1976; Bourne *et al.*, 1983; Kite, 1986; Hawksworth, this volume, Chapter 6). Plastids, and to a lesser extent mitochondria, as inhabitants-turned-organelles have retained a sufficient degree of identity to betray their origin even though they have partly relinquished by horizontal gene transfer (Sitte and Hansmann, 1986) the genetic control of some primary traits to the exhabitant. Beyond this point (where clues to the origin of, for example, microtubules may lie; Margulis and Bermudes, 1985) the evidence does not yet outweigh the scepticism. However, there are few theoretical objections to, and there is an increasing amount of supporting molecular evidence for, the contention that the ultimate potentially recognizable inhabitant may, like the lingering grin of the Cheshire cat (Smith, 1979), be a

DNA sequence spliced into the linear structure of the exhabitant's chromosomal DNA (Margulis, 1976, 1980).

Genetic integration need not be gradual and progressive. "Alien" DNA released during lysis of an inhabitant, intruder, or prey may accumulate in the exhabitant's nucleus, or be inserted by viral and plasmid vectors (Hartman, 1977). These "promiscuous genetic nomads", capable of bridging more casual cohabitants, are recognized to have been most effective in promoting coevolution within the bacterial "superorganism" (Reanney *et al.*, 1983; Sonea and Panisett, 1984).

Significant non-sexual transfer of novel genetic material in or between multicellular organisms is more difficult to visualize within the framework of traditional evolutionary, ecological and systematic thought, though the mechanism is increasingly being linked with macroevolutionary innovation (Went, 1971; Erwin and Valentine, 1984; Margulis and Bermudes, 1985). Eukaryotic genomes are being recognized as a mosaic of "fossilized" sequences "from both within the nucleus and without" (Lewin, 1983a; see also Marx, 1984; Bernardi *et al.*, 1985; Cherfas, 1985). The "without" embraces all of the biosphere. Cases for "gene jumps" between organisms as far removed phylogenetically as bacteria and sea urchins, angiosperms and man, and operating in either direction from prokaryote to eukaryote and *vice versa*, have been argued (cf. Lewin, 1982; Lamboy, 1984; Price *et al.*, 1986), and many puzzling phenomena, from the synthesis of morphine by plants to mammalian sex steroids by fungi (Kotala, 1984; Scott, 1985) beg a similar interpretation. "Increasingly we come to realize that multicellular organisms are not strictly the end products of a monophyletic line of DNA, but really complex organisms that have collected bits and pieces from different lines along the evolutionary road" (Southwood, 1985).

Plant–animal symbioses, traditionally viewed as one-to-one relationships, are, more often than not, mediated by vital members of the three microbial kingdoms (Monera, Protoctista, and Fungi) which, being phylogenetically much older, actively accompanied "higher" organisms throughout evolution (Southwood, 1985; Price *et al.*, 1986). Morphological, behavioural, chemical and genetic innovation can result from antagonistic or mutualistic symbiosis (Futuyma and Slatkin, 1983; Burdon, 1985; Price *et al.*, 1986). Undoubtedly, many such traits arise by a stepwise evolutionary or coevolutionary adjustment. However, the often intimate or protracted contact between microbial and plant or animal symbionts creates opportunities for the acquisition of novel genetic blueprints by uni- or bidirectional direct or plasmid-mediated transfer of genes between symbionts or integration of inhabitants. Furthermore, microorganisms capable of bridging distantly related lineages of potential symbionts may act as gene vectors in a manner suggested for virusus (Jeppsson, 1986).

Cecidia

"Rigid adherence to the ideas of natural selection has so far prevented a correct understanding of the significance of gall formation". Mani (1964, p. 225)

Some rainforest plants exhibit "nature's morphological caprices", which "the uninitiated would view with awe and take for malformations" (van Steenis, 1969). Conversely, among the multitude of plant galls, some could easily appear to the uninitiated to be normal plant components (Slepyan, 1961). Galls may resemble tubers, corms, bulbs, flowers, and fruits. Sometimes they appear to mimic organs of the host, for example, tuber-like galls of root nematodes on the Solanaceae, or pome-like galls of cinipid wasps on Rosaceae (Mani, 1964; Stone, 1979; Rohfritsch and Shorthouse, 1982). Sometimes galls resemble organs of non-host plants or those of alternative hosts. *Rhopalomyia* deforms leaves of *Aster scaber* into a structure unknown in asters but normal in other plants. Similarly, *Dasyneura* on *Plectranthus incisus* induces, as a secondary effect, the formation of an outgrowth that on the "correct" plant would constitute a normal "spur" (Mamaev, 1975, p. 220). Some oak midge galls bear stellate hairs that are otherwise unknown in oaks but are commonly produced in other groups of angiosperms (K. M. Harris, personal communication). Cinipid wasps infesting *Malus* and other Rosaceae and oaks reproduce "apples" on the latter.

Cecidial modification of flowers and fruits "often embraces an abnormal accentuation of these normal patterns" (Mani, 1964, p. 136). Modifications of androecia and gynoecia, which are usually confined to one of these organs, merely reproduce the "normal" condition, such as fusion of stamen filaments or swelling and coloration of the ovary wall. Sometimes galls are induced on more distant organs. Abnormalities of the inflorescence may result from the cecidogenous activity of root- or stem-gall symbionts (Mani, 1964), recalling the "secondary effects" in crown gall symbiosis, which may include epinasty* and the proliferation of adventitious roots (White, 1951).

Typically, however, the plant is provoked into surrounding the feeding site of the cecidozoan with highly nutritious cells. Cecidomyiid galls, for example, are rich in sugars. The concentration of highly assimilable nutrients makes galls vulnerable to other herbivores, but this is offset by the deposition of tannins, callose, cutin, or lignin that the cecidozoan induces, and which extends further the similarity of galls to plant organs, especially fruits (Fagan, 1918; Janzen, 1977, 1979). Slepyan (1961) claimed that the analogy extended not only to the superficial similarity between galls and fruits and seeds in architecture and histopography but also in their obeying the same principles of ripening. Also, according to Mani (1964, p. 209), in dehiscent

*Down-curving caused by more active growth on the upper side.

galls the specific mechanisms of opening are "not different from other dehiscent structures like fruits". The phenology of galls is linked to the life cycle of the cecidozoan: some complete it in a year or part of a year; others go through a supra-annual cycle of sexual and parthenogenetic phases. When galls fall, how they fall and how long they stay on plants provide further examples of their striking similarity to fruits. Rohfritsch and Shorthouse (1982) suggest that "we should resist referring to insect-induced galls as abnormal growth and consider them normal patterns of growth which are expressed once appropriate messages are applied", implicit in the nutritional behaviour and life cycle of the causal agent. Alternatively, the diversity of phylogenetically independent galls appears to be "normal" because an equally polyphyletic assortment of plant organs that they resemble *were* galls, the cecidogenous element of which is now encoded in the plant's genome and which evolution has fine-tuned for a different purpose.

Margulis and Bermudes (1988) reiterated Wallin's (1927) claim that symbiont-related hypertrophy may be an essential factor in tissue differentiation. Hyperplasia of reproductive parts, induced directly or indirectly by one cecidozoan but attracting phytophagous opportunists, could be adaptive if the plant benefited from a more effective pollination or dispersal of seed. Therefore, the consequences of the initial catastrophic stimulus of cecidial symbiosis should perhaps be included among examples of macroevolution by the intensification of the function of a pre-existing structure through intensification of selection pressure (Mayr, 1970). Cecidial modifications of growing apices and flower buds often involve condensation and simplification respectively. "Honey-dew" galls produce sugary exudates attracting nectar-feeding, and nectar-collecting insects, including honey bees. According to Iltis (1983), the angiospermous flower could have evolved by the telescoping of gynoecia "into the tightly focused reproductive efforts that animal-pollinated flowers represent and demand". The magnolioid reproductive structure, which dominates the early fossil record of angiosperms, is plastic as to the arrangement of parts, the variation involving various degrees of condensation and simplification (Endress, 1987). In this respect they are reminiscent of carpoid galls, the similarities extending to seeds that may be fleshy and conspicuously coloured (Slepyan, 1961). Succulent carpoid galls, which resemble fruits in form, smell, taste, and colour, are regularly eaten by frugivorous birds and mammals. Since this action is known to destroy cecidozoa rather than disperse them, obviously galls do not mimic fruits but could, more probably, be their antecedents.

A comparative study of present-day galls and fruits led Slepyan (1961) to conclude that during the evolution of galls the cecidozoans left increasing control of gall neoplasia to plants. Conversely, the formation of some fruits, such as the fig syconium, still requires an input from a cecidozoan. If the process is gradual, the cecidozoan at some point will become redundant and disassociated. The question then arises whether the two sets of observations are parts of the same evolutionary transition from a symbiont-induced neoplasm to a plant organ. Perhaps extant ant-plants provide the answer. Mani (1964, pp. 97, 240) claimed that stipular thorns of some African and Indian species of *Acacia* are the cecidia of gall-midges. Spent galls are usually colonized by ants. Of the diverse *successori* that inhabit empty galls generally, ants are by far the most important. Hocking (1970), on the other hand, considered the swollen stipular thorns of African acacias to be the plant's organs, albeit derived from homopteran, possibly aphid galls by "genetic assimilation". In the New World ant-acacias, no intermediaries are evident (Janzen, 1966): swollen thorns occur whether ants are present or not. Other traits of ant-plants—hollow stems, domatia, food-bodies and nectaries—are also produced irrespective of the presence of ants (Buckley, 1982). However, in the *Piper cenocladum–Pheridole bicornis* association, when ants are removed the production of food bodies nearly ceases only to resume when ants are reintroduced (Risch and Rickson, 1981). Here the food-bodies behave almost as if they were galls. Indeed, at least in the Müllerian bodies of *Cercopia peltata* their animal connection is suggested by the presence of glycogen (Rickson, 1971), which, to Margulis (1976) exemplified a plant product modified by a heterotrophic symbiont, though not necessarily by horizontal gene transfer. However, the Beltian bodies of some ant-acacias superficially resemble rhizobial, plasmid-induced stem nodules on the same family of plants (Legocki and Szalay, 1984).

Parallel or convergent evolution, such as that seen in fruits, can result from several causes, including one-time horizontal transfer of genes, if their phenotypic expression in the recipients resembles that in the donors (Went, 1971; Jeppsson, 1986). Being an ant-acacia is a saltatory trait (a neoseme *sensu* Margulis and Bermudes, 1985) that can be transmitted to non-ant-acacias as though it was controlled by a gene or a group of closely linked genes (Janzen, 1981). How strongly adaptive a neoseme has been may be judged from the extent of its spread, that is, by its weight as a taxonomic character. The taxonomic value of fruit fleshiness approaches zero.

Of immediate pertinence is the unanswered question of how cecidozoans "solicit genetic expression of plants" (Rohfritsch and Shorthouse, 1982). The "appropriate messages" are thought to be chemical. However, an individual host's response to a particular stimulus may vary from the formation of a typical gall to the elimination of the cecidozoan by necrosis. Further, sexual and parthenogenetic individuals of some species induce different galls on the same host species. Consequently, Mani (1964, p. 268) supposed that some combination of plant growth regulators and DNA might be involved in cecidogenesis.

Root-gall nematodes reduce or arrest meristematic growth and induce hyperplasia of the surrounding parenchyma by discharging unknown substances during feeding. The accompanying increase in host nucleic acids led Cohn (1975) to suggest that the rapid changes within the plant tissue "may be initiated by nucleic acids, presumably RNA, or their viral precursors injected by the nematode". Similarly, an increase in host nuclear RNA and DNA accompanies the proliferation of plasmodia of *Plasmodiophora brassicae* in gall tissue (Williams *et al.*, 1973). Proliferation of genetic elements in plants need not result from symbiosis, let alone gene transfer, though "theoretically the galls formed on plants under the influence of insects could also conceivably involve gene transfer mechanisms" (Zambryski *et al.*, 1983).

Margulis (1976) asked "Is the formation of highly specific colorful and differentiated insect galls on a single oak tree correlated with the transfer of genetic information from the insect into the plant tissue?" The answer, she suggested, would be found by searching for foreign genomes in insect galls; so far no one has. Classifications of galls, like those of flowers, fruits, and seeds are morphological or ecological; there are no genetic data on the origin and relationships of the basic morphological categories. Nevertheless, there is an amazing parallel between the history of the unravelling of the "tumorigenic principle" of crown gall and the current speculation on the identity of the cecidogenous "factor x", whether it is called "some biologically active substance" injected by the cecidozoan, or, more daringly, DNA.

Speculation on the mechanisms of cecidial symbiosis must include a potential complication that concerns the origin of the "cecidogenous factor" in gall-inducing animals. Phytophagy and herbivory are, more often than not, mediated by microorganisms (e.g., gut symbionts), and many cecidial symbioses involve microbial intermediaries that probably represent an ancestral trait (Bissett and Borkent, this volume, Chapter 9; Mamaev, 1975). Some microbial symbionts of insects have increasingly become subordinated as inmates of mycetocytes, perhaps even further as in some other "Blochmann bodies". A more intimate integration might easily escape detection: supernumerary genomes are most visible only when they elicit what we perceive, from the perspective of little more than a human life-span, as abnormalities in the host.

Mycocecidia

"The study of the mechanisms of the development of galls and terata may play an important role in understanding the nature of seeds and fruits". Slepyan (1961)

Fungi form intimate, stable, and prolonged associations with plants. The intimacy of the association frequently results from the facility with which fungi enter plant cells by the mechanical or enzymatic breaching of cell walls and the modification of membranes to short-circuit organic carbon directly from the source. They also demonstrate a diversity of extranuclear, cytoplasmically inherited characteristics that may be associated with viruses or plasmids (Garber *et al.*, 1986) and act as intragenomic vectors of genetic blueprints for the synthesis of novel metabolites (Hamill *et al.*, 1987).

Within the fungi, Gaber and Leonard (1981) demonstrated unilateral gene transfer between nuclei in a dikaryon of *Schizophyllum commune*; Setliff (1983) invoked horizontal gene transfer between a holo- and a phragmobasidiomycete, presumably via hyphal anastomoses, to account for the "hybrid" nature of *Aporium*; and if ancestral fungi engaged in nuclear swappings such as have been demonstrated in extant red algae (Goff and Coleman, 1984), some such event would go a long way towards explaining the evolution of the complex mating systems of dikaryomycetes that are so difficult to interpret in terms of gradualistic natural selection. "The probable or even possible evolutionary history of tetrapolar incompatibility is completely obscure, and the selective advantages of the system, though qualitatively obvious are quantitatively puzzling" (Raper, 1968).

Lamboy (1984), who explicitly hypothesized that horizontal gene transfer from fungi was a significant aspect of the evolution of flowering plants, considered that fungi met all the conditions necessary for a direct gene transfer without a viral intermediary, and for serving as vectors of plant and animals genes. Plant nuclei readily integrate foreign DNA (Leber and Hemleben, 1983), even when injected mechanically (Klein *et al.*, 1987). Yet, as Lamboy admitted, not a single documented example of horizontal gene transfer between fungi and plants is known, though this may simply be because no one has looked. Many, however, have speculated. Ragan and Chapman (1978, p. 127) hinted at the possibility of horizontal gene transfer from ligneous plants to fungi to account for the presence of lignin biodegrading enzymes in wood-rotting dikaryomycetes. Atsatt (1973, 1986) hypothesized that the haustoria of parasitic plants evolved, repeatedly and independently of habitat, from neoplasms induced by inherited but as yet undetected endophytes, possibly bacteria carried by a fungus (Price *et al.*, 1986). Haustorial invasion by *Cuscuta* occasionally induces tumours in the host (Dean, 1937), significantly sometimes as "secondaries" reminiscent of those of crown gall. The formation of a direct vascular connection between a plant host and its parasite, like vascularization in some cecidia, could result from microbial action. Lamboy (1984) compared the differentiation of vessels in angiosperms to some consequences of fungal attack. Pirozynski (1981) suggested that vascularization and lignification of tracheophyte precursors could have been the product of mycotrophic symbiosis. The case for horizontal gene transfer from fungi to plants may, however, be emerging from the realm of anecdote: the current research on fungal trichothecene toxins or their derivatives in two Brazilian species of *Baccharis* (Jarvis *et al.*, 1987) promises to supply the empirical proof (B. B. Jarvis, personal communication).

Fungus-induced neoplasms vary greatly in form and organization. Many examples of "organoid mycocecidia" (Mani, 1964), such as fasciations, witches' brooms, or root hyperplasias, show sufficient similarity to both zoocecidia and such normal features of plants as virescences, synanthy, or food-storing roots to cause one to suspect that the cecidogenous "factor x" could be a fungus or could have been acquired or derived from fungi. *Taphrina*, which causes "leaf curl" hyperplasia, is a common inhabitant of insect guts and mycetomes (Jurzitza, 1979), and a favoured insect food: in British Columbia, *Taphrina* spots on fronds of sword fern are regularly eaten out (Pirozynski, unpublished observation).

Slepyan (1961) drew attention to data indicating that the processes of induction of parthenocarpic fruit in nature or laboratory, the encapsulation of the embryo by proliferating ovary cells in apomictic and sexual fruits, and the encapsulation in plant tissues of a cecidozoan, all involve the same growth regulators; only their induction sites or agencies change.

Plant tumours and their enhanced nutrient content are correlated with increased concentration of growth regulators (Kado, 1984; Goodman *et al.*, 1986). Tumour-inducing *Agrobacterium*, *Corynebacterium*, *Pseudomonas*, *Plasmodiophora*, *Taphrina*, *Melampsora*, or *Ustilago* species free indole auxins in the host, stimulate the host to increase their production, or release their own, thereby creating an imbalance resulting in a neoplasm (Bailey, 1986). In *Agrobacterium tumefaciens*, genes in the *vir*-region of the Ti plasmid induce the production of abnormally high concentrations of IAA and cytokinin in the plant. In other tumour-inducing bacteria, and perhaps also the fungi, plant growth regulators appear to be produced by these organisms themselves; that is, they are encoded by chromosomal genes. However, at least in the "Oleander strain" of *Pseudomonas savastanoi* tumour-inducing IAA production is correlated with the presence of a plasmid (Comai and Kosuge, 1980). It would be helpful to understand why plant growth regulators occur in symbiotic bacteria and fungi. Suggestions that the genes encoding such substances in prokaryotes initially came from plants (Chilton, 1983; Tempé *et al.*, 1984) are not supported by the evidence from the Ti plasmid where these genes are located in the prokaryotic *vir*-region. Alternatively, their presence in plants may be a long-standing acquisition from microbes (Kado, 1984). Therefore, it is possible that some present-day tumorigens may only be recalling, by activation of genes, events from the long history of coevolution by "jumping genes".

Fungi are debatably the most important of the free-living, coevolved symbionts of zoocecidia (Mani, 1964, p. 231). An element of doubt may have been reinforced by Batra and Lichtwardt (1963), who considered typical fungal inhabitants of insect galls to be opportunistic *successori*, either as parasites on nutrient-enriched gall tissues or as saprobes on insect excreta. They further claimed that if the fungi were dispersed by insects, it was merely as passive external contaminants. However, recent findings (Borkent and Bissett, 1985; this volume, Chapter 9) refocus attention on "mycozoocecidia" as products of mutualistic symbiosis and coevolution, and on their wide representation in nature. In the *Asteromyia* blister gall on *Solidago*, a species of *Macrophoma* introduced as spores from mycangia at oviposition forms a characteristic stroma that houses the developing larva and protects it from parasitoids (Weis, 1982). As Batra (1964) pointed out, the association is no longer casual, because the stroma is formed only in the presence of larvae: the control of this characteristic fungal expression has been acquired by the insect (Mamaev, 1975). These findings also refocused attention on Mamaev's contention that in the evolution of gall midges (Cecidomyiidae) feeding adaptations progressed from a diet of detritus fungi to fungi growing necrotrophically on injured plants, to their gradual but complete replacement by plant sap. Considering that plant sap is notoriously poor and imbalanced as the sole diet and is normally exploited by symbiotic systems, a postulate disregarding symbiosis is an incomplete scenario.

It is possible that many fungi that are now specialized, "cultivated" mutualists of mycozoocecidia originated as opportunistic *successori*. It is just as probable that an earlier, concomitant or alternative association with fungi that induce neoplasms would prove equally advantageous. Furthermore, the trait selected for need not have depended on partial autonomy or even the physical presence of the fungus. The case in point may be the diverse, polyphyletic ant-associated epiphytes of which *Myrmecodia* and related genera of the Rubiaceae are the most bizarre. The ants live in a chambered "tuber" derived from the hypocotyl, keeping broods in chambers having smooth, nonabsorptive walls, and depositing faeces and organic debris in chambers with warted, absorptive walls. A specific hyphomycete is normally associated with each kind of chamber: an unnamed biotrophic species in the smooth, and a saprotrophic *Arthrocladium* in the warted. Little is known of their function, but both fungi appear to be dispersed by ants, may serve as the ants' food, and appear to play a role in the association of the ants with plants. More specifically, *Arthrocladium* is credited with rendering organic debris assimilable by the plant (Huxley, 1978, 1982). Neither of the two fungi is likely to be the primary agent of "tuberization", nor is it likely that the ants, like those of ant-acacias, are anything but *successori*, though they may have been responsible for these plants' epiphytic habit. Whatever stimulus led to the evolution of the morphological abnormality that is *Myrmecodia*, it is, presumably, no longer provided by autonomous symbionts: plants grown from seeds form "normal" tubers irrespective of whether ants and fungi are present.

SEX, PLASMIDS AND HISTOGENESIS

"Extrachromosomal DNA may well be the key for the understanding of communication at the genetic level between different organisms living together in a host–parasite or a symbiontic relation". Meinhardt (1984)

Sex

In the view of Sagan and Margulis (1987), the evolution of sex (meiosis) is an historical accident, the initial advantage of which was "relief" from the burden of supernumerary genomes in "cybrids" resulting from cannibalistic fusions between unicellular protoctists. The continuing success of sex may be because it achieves exactly that. The regulation of spontaneous "epigenetic defects" (Holliday, 1987) could be an acquired adaptation. When "sex became entangled with reproduction" (Sagan and Margulis, 1987), a basic mechanism accrued for the filtering out of "infection" acquired by the parents. In consequence, the Earth came to be populated by increasingly discrete, specific groups "of individuals whose common gene pool is protected against the inflow of alien genes" (Lewin, 1983b). Concomitantly, the symbiotic sharing of complementary genetic resources offered competitive advantage to stable mutualistic partnerships (Law, 1985). The direction of host–parasite coevolution (Price *et al.*, 1986) and of symbiosis generally is towards mutualism and ultimately stable integration into the genome of one partner of those genes of the other that give the partnership its competitive advantage. Perhaps sex and symbiosis may be viewed as opposing selective forces that drive the coevolution of sexuality and complexity. As the complexity of organisms increases, and with it the possibility of multiple levels of symbiotic interaction (Margulis, 1980), so increases their dependence on sexual reproduction (Wertheim, 1986). As pieces of alien genomes "filter" through the defences of sex, so increases the effectiveness of meiosis in neutralizing them by "repair" and "rejuvenation" of the native genome. And as alien genes bring blueprints for saltatory innovations, meiosis may well play an increasing role in their organization, and the "canalization" of neosemes (cf. Sagan and Margulis, 1987). This is because for those alien genes that succeeded in entering the host genome, sex provides a means for their unrestricted spread (Rose and Doolittle, 1983). However, before sex can act as friend it must be defeated as foe. In most symbiotic interactions, sex maintains an arm's length relationship between the participants. The advantages of such relationships derive from the participant's autonomy (i.e., reproductive isolation) as, for example, in ectomycorrhizas, pollination, or seed dispersal (Pirozynski and Malloch, this volume, Chapter 10). If coevolved mutualism ensues, it is more likely to result from indirect, reciprocal genetic adjustments.

In more protracted and intimate antagonistic symbiosis, the pressure from parasites—threat of genetic integration—provides a strong incentive for the maintenance of sex (Levin, 1975; Hamilton, 1980; Rice, 1983; Law, 1985; Lively, 1987; Clay, this volume, Chapter 4). Characteristically, in intimate mutualistic associations in which one partner has been stably integrated as a more or less intracellular inhabitant, the *inhabitant's* capacity for sexual reproduction has been suppressed or eliminated (Law and Lewis, 1983; Law, 1985; Hawksworth, this volume, Chapter 6). However, if stable, mutualistic endosymbiosis exposes the exhabitant's genome to the inhabitant's genes, the key step to successful integration should be suppression of the *exhabitant's* sexuality.

Plasmids

In many organisms, the expression of the sexual state (gender) as well as the adoption of a sexual or asexual mode of reproduction is induced by the outside environment (Policansky, 1987), including symbiosis that often results in the suppression of sexual reproduction in one or other cohabitant. Many symbiotic eukaryotes carry maternally transmitted microorganisms and viruses that suppress the production of males (Werren *et al.*, 1986). Male sterilization by parasites may lead to the evolution of parthenogenetic siblings (Price *et al.*, 1986), and exemplifies one aspect of parasite-induced incompatibility of populations that may lead to genetic isolation and speciation (Wade and Stevens, 1985; Thompson, 1986).

Environmentally induced sex-switches in plants may operate via growth regulators. For example, the modification of flowers of some grasses and sedges into viviparous vegetative plantlets can be artificially induced by altering the phytohormonal balance (Clay, 1986). Regarding the genetic basis of the interaction, little can be added to what has already been noted above (p. 257). It should, however, be recognized that extrachromosomal genes are increasingly being implicated in the generation of phenotypic cytoplasmic male sterility in plants, that is, a maternally inherited defect that impairs the production of viable pollen. An experimental transfer of organelles or organelle-controlled features from donor to recipient results in cybrid plants showing floral malfunctions and male sterility (Aviv *et al.*, 1984). "Spontaneous" cytoplasmic male sterility in plants appears to result from the synthesis of a novel chimaeric gene through rearrangements of nucleotides within the mitochondrial genome; deletion of this gene restores fertility (Butow, 1986). Genetic rearrangements responsible for the loss and restoration of fertility (sex) may be, or may have been, mediated by plasmids: the sterility/fertility traits that correlate with the gain or loss of mitochondrial (mt) plasmids, or the genes responsible for these traits, may have been transferred and incorporated into the nucleus (Levings *et al.*, 1980; Tudzynski *et al.*, 1986). Extrachromosomal mt plasmids are a ubiquitous phenomenon: in most of the higher plants investigated, mt linear plasmids associated with the cytoplasmic male sterility syndrome obviously "belong to the normal complement of the chondriome" (Tudzynski *et al.*, 1986).

Several fungal pathogens are "sterilizing", producing more persistent and robust "eunuch" plants. Gall-inducing fungi, and to a lesser extent the cecidozoa, can suppress sex or reverse sexuality, sometimes as "indirect castration" away from the site of the interaction. Remarkable cases of sex-reversal accompany mycocecidia of *Ustilago* and some other smuts of flowers. If galls are normally found on staminate flowers, a chance infection of a pistillate flower will transform it into a staminate flower and vice versa (Mani, 1964, p. 252). A systemic infection of some grasses and sedges with endophytic clavicipitaceous fungi may result in the modification of flowers into viviparous vegetative plantlets (Clay, 1986).

The discoveries of extrachromosomal, cytoplasmically inherited genetic elements came relatively late in fungi but they now include a variety of DNA and RNA viruses, plasmids, and "D-factors" (Esser *et al.*, 1983; Garber *et al.*, 1986; Rogers *et al.*, 1986). The role, especially of mt plasmids, in fungi is still largely unknown (Nargang, 1985), but indications are that viruses and plasmids may be involved with expressions of virulence and sexual transmutations in fungus–plant interactions: hypovirulence of *Rhizoctonia*, virulence of *Gaeumannomyces graminis* (Meinhardt, 1984), suppressed fertility of *Cochliobolus* (Nargang, 1985), or production of "killer" toxins in *Ustilago maydis* and yeasts. Particularly exciting are preliminary claims that plasmids of some pathogenic fungi, especially those in clavicipitaceous "sterilizing" endophytes, show some homology with the S plasmids that correlate with cytoplasmic male sterility in corn and possibly in other usual graminaceous hosts of these fungi (Esser *et al.*, 1983; Tudzynski *et al.*, 1986).

Histogenesis

Successful experiments involving the suppression of sex are difficult to visualize in the anthropocentric context of mammalian hegemony, even though clonality in animals generally is a trait that facilitates the evolution of mutualism (Wulff, 1985). Applied to higher vertebrates, Maynard-Smith's (1984) statement that "the abandonment of sex condemns a species to an early extinction" is an understatement. Yet, as already noted, among the invertebrates and certainly within plants, symbiotic integration often results in the suppression of sexual reproduction in the host. Here, abandonment of sex condemns a species to a quantum evolutionary change. From the perspective of geological time it is obligate sexuality that signals an evolutionary dead-end and a short-cut to extinction.

Clay (this volume, Chapter 4) demonstrated that symbioses between clavicipitaceous endophytic fungi and their plant hosts form a continuum having an antagonistic relationship in which the host is sterilized by a sexual parasite at one end, and at the other a mutualistic relationship in which the host is sexual but the parasite loses sex. Sterilization of the host permits or promotes greater integration of the symbiont, since the evolution of novel organs of vegetative propagation is to propagate "*infected*" plants. During a subsequent reversal to sexuality, plants not only regain "fitness" but gain the advantage of an anti-herbivore mutualist (or rather its active metabolites), now rendered asexual and maternally transmitted (seed-borne). The advantages in mutualistic symbioses of the inhabitant's loss of sexual recombination as the source of destabilizing change has been discussed elsewhere (Law and Lewis, 1983; Rice, 1983; Law, 1985).

The macroevolutionary implications of symbiont-induced sexual transmutation with concomitant neoplasia for the origins of various organs of vegetative propagation and perennation, of biennial and perennial life cycles in primarily annual plants, and for histogenesis generally, remains to be explored. In the case of grasses and sedges, Clay (this volume, Chapter 4) demonstrated that infection by clavicipitaceous endophytes promoted the survival of viviparous plantlets and tubers. Could it be that plants that switch gender "at will", such as dwarf ginseng (Schlessmann, 1987) or Jack-in-the pulpit (Policansky, 1987), have undergone symbiont-induced sexual transmutation betrayed in dwarfed habit and the presence of storage organs as the means of vegetative propagation? Other examples of plants whose sexuality may have been tampered with by symbionts are amphicarpic plants that produce both aerial and subterranean flowers and seeds. The adaptation offers several ecological advantages, and can be linked with the evolution of lineages of cleistogamy, seed dimorphism, and dual strategies of seed dispersal (Cheplick, 1987). What is more difficult to visualize within the framework of gradual adaptive change is how the trait has originated.

Silander (1985) claimed that "Lamarckian effects", whereby information passes from the environment to the DNA, are most likely to be seen in clonal plants. In this scenario, environmentally induced somatic changes could become incorporated in the germline by horizontal gene transfer. Because the most immediate environment of plants consists of microbial symbionts, the incorporation of alien genes by the same mechanism could equally occur. In both situations plasmids may come to play a crucial role.

Atsatt (1986), who hypothesized the origin of haustoria in plant-parasitic angiosperms from tumours induced by microbes, considered that stable integration of tumorigenic microbial genes in the host plant genome was a real possibility. The relatively recent derivation of the maize ear from teosinte tassel was claimed to have been through catastrophic transmutation involving condensation and feminization by growth substances, possibly in conjunction with viral or fungal infection (Iltis, 1983). More specifically, Lamboy (1984) postulated a horizontal gene transfer from corn smut, *Ustilago maydis*. A comparable case of quantum evolutionary change, but in an earlier stage of canalization, may be the horticultural induction by *Ustilago esculenta* of succulence in vegetative parts of plants used as vegetables in China. The speed with which such abnormalities are canalized by subsequent selective breeding by man indicates that similar natural "catastrophes" could almost certainly be canalized by millions of years of natural selection.

ACKNOWLEDGEMENTS

Several colleagues, namely Dr A. Borkent, Dr J. Gavora, Professor D. L. Hawksworth, Professor D. H. Lewis, Professor L. Margulis, Dr D. B. O. Savile, and Dr D. T. Wicklow commented on earlier drafts or directed me to relevant literature. Their advice, although not always taken, and their help, are gratefully acknowledged. I am particularly indebted to Professor D. H. Lewis for stimulating discussions that provided ideas and incentive for this compilation.

REFERENCES

Agayev, M. G. (1968). [On the versatility of speciation process.] *Bot. Zh. SSSR* **53,** 23–38.

Atsatt, P. R. (1973). Parasitic flowering plants: how did they evolve? *Am. Nat.* **107,** 502–510.

Atsatt, P. R. (1986). Evolution of parasitic plants: are haustoria modified tumors? In "Parasitic Weeds in Agriculture" (L. Musselman, ed), vol. 1, pp. 127–141. CRC Press, Boca Raton, Fla.

Aviv, D., Arzee-Gonen, P., Bleichman, S., and Galun, E. (1984). Novel alloplasmic *Nicotiana* plants by "donor-recipient" protoplast fusion: cybrids having *N. tabacum* or *N. sylvestris* nuclear genomes and either or both plastomes and chondriomes from alien species. *Molec. gen. Genet.* **196,** 244–253.

Bailey, J. A. (ed) (1986). "Biology and Molecular Biology of Plant–Pathogen Interactions." Springer, Berlin.

Batra, L. R. (1964). Insect–fungus blister galls on *Solidago* and *Aster*. *J. Kans. ent. Soc.* **37,** 227–234.

Batra, L. R. and Lichtwardt, R. W. (1963). Association of fungi with some insect galls. *J. Kans. ent. Soc.* **36,** 262–278.

Bermudes, D. and Margulis, L. (1987). Symbiont acquisition as neoseme: origin of species and higher taxa. *Symbiosis* **4,** 185–198.

Bernardi, G., Oloffson, B., Filipski, J., Zehial, M., Selinas, J., Cuny, G., Meunier-Rotival, M., and Rodier, F. (1985). The mosaic genome of warm-blooded vertebrates. *Science, N.Y.* **288,** 953–958.

Borkent, A. and Bissett, J. (1985). Gall midges (Diptera: Cecidomyiidae) as vectors for their fungal symbionts. *Symbiosis* **1,** 185–194.

Bourne, G. H., Danielli, J. F., and Jeon, K. W. (eds) (1983). Intracellular symbiosis. *Internat. Rev. Cytol., Suppl.* **14,** 1–441.

Buchanan-Wollaston, V., Passiatore, J. E., and Cannon, F. (1987). The *mob* and *ori T* mobilization functions of bacterial plasmid promote its transfer to plants. *Nature, Lond.* **328,** 172–175.

Buckley, R. C. (1982). Ant–plant interactions: a world review. In "Ant–Plant Interactions in Australia" (R. C. Buckley, ed), pp. 111–162. W. Junk, The Hague.

Burdon, J. J. (1985). Pathogens and genetic structure of plant populations. In "Studies on Plant Demography" (J. White, ed), pp. 313–325. Academic Press, London.

Butow, R. A. (1986). Rearranging the plant genome. *Nature, Lond.* **324,** 620.

Cherfas, J. (1985). When is a tree more than a tree? *New Scientist* **106**(1461), 42–45.

Cheplick, G. P. (1987). The ecology of amphicarpic plants. *Trends Ecol. Evol.* **2,** 97–101.

Chilton, M.-D. (1983). A vector for introducing new genes into plants. *Scient. Am.*, **1983** (June), 51–59.

Chilton, M.-D., Tepfer, D. A., Petit, A., David, C., Casse-Delbart, F., and Tempé, J. (1982). *Agrobacterium rhizogenes* inserts T-DNA into the genomes of the host plant root cells. *Nature, Lond.* **295,** 432–434.

Clay, K. (1986). Induced vivipary in the sedge *Cyperus virens* and the transmission of the fungus *Balansia cyperi* (Clavicipitaceae). *Can. J. Bot.* **64,** 2984–2988.

Cohn, E. (1975). Relations between *Xiphinema* and *Longidorus* and their host plants. In "Nematode Vectors of Plant Viruses" (F. Lamberti, C. E. Taylor, and J. W. Seinhorst, eds), pp. 365–384. Plenum Press, London and New York.

Comai, L. and Kosuge, T. (1980). Involvement of plasmid deoxyribonucleic acid in indoleacetic acid synthesis in *Pseudomonas sevastanoi*. *J. Bact.* **143,** 950–957.

Dean, H. L. (1937). Gall formation in host plants following haustorial invasion by *Cuscuta*. *Am. J. Bot.* **24,** 167–173.

Endress, P. K. (1987). The early evolution of the angiosperm flower. *Trends Ecol. Evol.* **2,** 300–304.

Erwin, D. H. and Valentine, J. W. (1984). "Hopeful monsters", transposons, and metazoan radiation. *Proc. natn. Acad. Sci. U.S.A.* **81,** 5482–5483.

Esser, K., Kück, U., Stahl, U., and Tudzynski, P. (1983). Cloning vectors of mitochondrial origin in eukaryotes: a new concept in genetic engineering. *Curr. Genet.* **7,** 239–243.

Fagan, M. M. (1918). The uses of insect galls. *Am. Nat.* **52,** 155–176.

Futuyma, D. J. and Slatkin, M. (eds) (1983). "Coevolution." Sinauer Associates, Sunderland, Mass.

Gaber, R. F. and Leonard, T. J. (1981). Unilateral internuclear gene transfer and cell differentiation in *Schizophyllum*. *Nature, Lond.* **291,** 342–344.

Garber, R. C., Lin, J. J., and Yoder, O. C . (1986). Mitochondrial plasmids in *Cochliobolus heterostrophus*. In "Extrachromosomal Elements in Lower Eukaryotes" (R. B. Wickner, A. Hinnebush and A. M. Lambowitz, eds), pp. 105–118. Plenum Press, New York and London.

Goff, L. J. and Coleman, A. W. (1984). Transfer of nuclei from a parasite to its host. *Proc. natn. Acad. Sci. U.S.A.* **81,** 5420–5424.

Goodman, R. N., Király, Z., and Wood, R. K. (1986). "The Biochemistry and Physiology of Plant Disease." University of Missouri Press, Columbia.

Hamill, J. D., Parr, A. J., Rhodes, M. J. C., Robins, R. J., and Walton, N. J. (1987). New routes to plant secondary products. *Bio/Technology* **5,** 800–804.

Hamilton, W. D. (1980). Sex versus non-sex versus parasite. *Oikos* **35,** 282–290.

Hartman, H. (1977). Speculation on viruses, cells and evolution. *Evol. Theor.* **3,** 159–163.

Hepburn, A. (1983). Mother nature got there first. *New Scientist* **99**(1377), 923–925.

Hocking, B. (1970). Insect associations with swollen thorn acacias. *Trans. R. ent. Soc. Lond.* **122,** 211–255.

Holliday, R. (1987). The inheritance of epigenetic defects. *Science, N.Y.* **238,** 163–170.

Horsch, R. B., Fraley, R. T., Rogers, S. R., Sanders, P. R., Lloyd, A., and Hoffmann, N. (1984). Inheritance of functional foreign genes in plants. *Science, N.Y.* **233,** 496–498.

Huxley, C. R. (1978). The ant-plants *Myrmecodia* and *Hydnophytum* (Rubiaceae), and the relationships between their morphology, ant occupants, physiology and ecology. *New Phytol.* **80,** 231–268.

Huxley, C. R. (1982). Ant-epiphytes in Australia. In "Ant–Plant Interactions in Australia" (R. C. Buckley, ed), pp. 63–73. W. Junk, The Hague.

Iltis, H. H. (1983). From teosinte to maize: the catastrophic sexual transmutation. *Science, N.Y.* **222,** 886–894.

Janzen, D. H. (1966). Coevolution of mutualism between ants and acacias in Central America. *Evolution* **20,** 249–275.

Janzen, D. H. (1977). Why fruits rot, seeds mould, and meat spoils. *Am. Nat.* **111,** 691–713.

Janzen, D. H. (1979). Why food rots. *Nat. Hist., N.Y.* **88,** 60–64.

Janzen, D. H. (1981). The defenses of legumes against herbivores. In "Advances in Legume Systematics" (R. M. Polhill and P. H. Raven, eds), vol. 2, pp. 951–977. Royal Botanic Gardens, Kew.

Jarvis, B. B., Wells, K. M., Lee, Y.-W., Bean, G. A., Kommedahl, T., Barros, C. S., and Barros, S. S. (1987). Macrocyclic trichothecene mycotoxins in Brazilian species of *Baccharis. Phytopathology* **77,** 980–984.

Jeon, K. W. and Danielli, J. F. (1971). Microsurgical studies with large free living amoebas. *Int. Rev. Cytol.* **30,** 49–89.

Jeon, K. W. and Jeon, M. S. (1976). Endosymbiosis in amoeba: recently established endosymbionts have become required cytoplasmic components. *J. Cell Physiol.* **89,** 337–344.

Jeppsson, L. (1986). A possible mechanism in convergent evolution. *Paleobiology* **12,** 80–88.

Jurzitza, G. (1979). The fungi symbiotic with anobiid beetles. In "Insect–Fungus Symbiosis" (L. R. Batra, ed), pp. 65–76. Wiley, New York.

Kado, C. I. (1984). Phytohormone-mediated tumorigenesis by plant pathogenic bacteria. In "Genes Involved in Microbe–Plant Interactions" (D. P. S. Verma and Th. Hohn, eds), pp. 311–336. Springer-Verlag, Vienna and New York.

Kite, G. (1986). Evolution by endosymbiosis: the inside story. *New Scientist* **111**(1515), 50–52.

Klein, T. M., Wolf, E. D., Wu, R., and Sanford, J. C. (1987). High-velocity microprojectiles for delivering nucleic acids into living cells. *Nature, Lond.* **327,** 70–73.

Kotala, G. (1984). Steroid hormone systems found in yeast. *Science, N.Y.* **225,** 913–914.

Krassilov, V. A. (1973). Mesozoic plants and the problem of angiosperm ancestry. *Lethaia* **6,** 163–178.

Krassilov, V. A. (1977). The origin of angiosperms. *Bot. Rev.* **43,** 143–176.

Lamboy, W. T. (1984). Evolution of flowering plants by fungus-to-host horizontal gene transfer. *Evol. Theor.* **7,** 45–51.

Law, R. (1985). Evolution in mutualistic environment. In "The Biology of Mutualism" (D. H. Boucher, ed), pp. 145–170. Croom-Helm, London.

Law, R. and Lewis, D. H. (1983). Biotic environments and the maintenance of sex—some evidence from mutualistic symbioses. *Biol. J. Linn. Soc.* **20,** 249–276.

Leber, B. and Hemleben, V. (1983). Uptake of homologous DNA into nuclei of seedlings and by isolated nuclei of higher plant. *Z. PflPhys.* **91,** 305–316.

Legocki, R. P. and Szalay, A. A. (1984). Molecular biology of stem nodulation. In "Genes Involved in Microbe–Plant Interactions" (D. P. S. Verma and Th. Hohn, eds), pp. 225–268. Springer, Vienna and New York.

Levin, D. A. (1975). Pest pressure and recombination system in plants. *Am. Nat.* **109,** 437–451.

Levings, C. S., Kim, B. D., Pring, D. R., Conde, M. F., Mans, R. J., Laughnan, J. R., and Gabay-Laughnan, S. J. (1980). Cytoplasmic reversion of CMS-S in maize: association with a transpositional event. *Science, N.Y.* **209,** 1021–1023.

Lewin, R. (1982). Can genes jump between eukaryotic species? *Science, N.Y.* **217,** 42–43.

Lewin, R. (1983a). Promiscuous DNA leaps all barriers. *Science, N.Y.* **219,** 478–479.

Lewin, R. (1983b). Invasion by alien genes. *Science, N.Y.* **220,** 811.

Lively, C. M. (1987). Evidence from a New Zealand snail for the maintenance of sex by parasitism. *Nature, Lond.* **328,** 519–521.

Mamaev, B. M. (1975). "Evolution of Gall-Forming Insects: Gall Midges." (Transl. A. Crozy.) British Library Board, Boston Spa.

Mani, M. S. (1964). "Ecology of Plant Galls." W. Junk, The Hague.

Margulis, L. (1976). Genetic and evolutionary consequences of symbiosis. *Expl. Parasit.* **39,** 277–349.

Margulis, L. (1980). Symbiosis and parasexuality. In "Cellular Interactions in Symbiosis and Parasitism" (C. B. Cook, P. W. Pappas, and E. D. Rudolph, eds), pp. 263–273. Ohio State University Press, Columbus.

Margulis, L. and Bermudes, D. (1985). Symbiosis as a mechanism of evolution: status of cell symbiosis theory. *Symbiosis* **1,** 101–124.

Margulis, L. and Bermudes, D. (1988). Symbiosis and evolution: a brief guide to recent literature. In "Cell-to-cell Signals in Plant–Animal and Microbial Symbiosis" (S. Scannerini, D. C. Smith, P. Bonfante and V. Gianinazzi-Pearson, eds). Springer, Berlin (in press).

Marx, J. L. (1984). Instability in plants and the ghost of Lamarck. *Science, N.Y.* **224,** 1415–1416.

Marx, J. L. (1985). How Rhizobia and legumes get it together. *Science, N.Y.* **230,** 157–158.

Maynard-Smith, J. (1984). The ecology of sex. In "Behavioural Ecology" (J. R. Krebs and N. B. Davies, eds), pp. 201–221. Blackwell Scientific Publications, Oxford.

Mayr, E. (1970). "Populations, Societies and Evolution." Belknap Press, Cambridge, Mass.

Meinhardt, F. (1984). Biological control of plant pathogenic fungi. *Progr. Bot.* **46,** 241–247.

Nargang, F. E. (1985). Fungal mitochondrial plasmids. *Expl. Mycol.* **9,** 285–293.

Pirozynski, K. A. (1981). Interactions between fungi and plants through the ages. *Can. J. Bot.* **59,** 1824–1827.

Policansky, D. (1987). Sex choice and reproductive costs in jack-in-the-pulpit. *BioScience* **37,** 476–481.

Price, P. W., Westoby, M., Rice, B., Atsatt, P. R., Fritz, R. S., Thompson, J. N., and Mobley, K. (1986). Parasite mediation in ecological interactions. *A. Rev. ecol. Syst.* **17,** 487–505.

Ragan, M. A. and Chapman, D. J. (1978). "A Biochemical Phylogeny of the Protists." Academic Press, New York.

Raper, J. R. (1968). On the evolution of fungi. In "The Fungi" (G. C. Ainsworth and A. S. Sussman, eds), Vol. 3, pp. 677–690, Academic Press, New York and London.

Reanney, D. C., Gowland, P. C., and Slater, J. H. (1983). Genetic interactions among microbial communities. In "Microbes in Their Natural Environments" (J. H. Slater, R. Whittenbury, and J. W. T. Wimpenny, eds), pp. 379–421. Cambridge University Press, Cambridge and London.

Rice, W. R. (1983). Parent-offspring pathogen transmission: a selective agent promoting sexual reproduction. *Am. Nat.* **121,** 187–203.

Rickson, F. R. (1971). Glycogen plastids in Müllerian body cells of *Cecropia peltata*—a higher green plant. *Science, N.Y.* **173,** 344–347.

Risch, S. J. and Rickson, F. R. (1981). Mutualisms in which ants must be present before plants produce fruit bodies. *Nature, Lond.* **291,** 149–150.

Rogers, H. J., Buck, K. W., and Brasier, C. M. (1986). The D2-factor in *Ophiostoma ulmi*: expression and latency. In "Biology and Molecular Biology of Plant-Pathogen Interactions" (J. A. Bailey, ed), pp. 393–400. Springer, Berlin.

Rohfritsch, O. and Shorthouse, J. D. (1982). Insect galls. In "Molecular Biology of Plant Tumors" (G. Kahl and J. S. Schell, eds), pp. 131–152. Academic Press, New York.

Rose, R. and Doolittle, F. (1983). Parasitic DNA—the origin of species and sex. *New Scientist* **98**(1362), 787–789.

Sagan, D. and Margulis, L. (1987). Cannibals' relief: the origins of sex. *New Scientist* **115**(1572), 36–39.

Schlessman, M. A. (1987). Gender modification in North American ginsengs. *BioScience* **37,** 469–475.

Scott, A. (1985). Messages in evolution. *New Scientist* **107**(1466), 30–31.

Setliff, E. C. (1983). *Aporium*—an example of horizontal gene transfer? *Mycotaxon* **18,** 19–21.

Silander, J. A. (1985). Microevolution in clonal plants. In "Population Biology and Evolution of Clonal Organisms" (J. B. C. Jackson, L. W. Buss, and R. E. Cook, eds), pp. 107–152. Yale University Press, New Haven, Conn.

Sitte, P. and Hansmann, P. (1986). Cytosymbiosis. *Progr. Bot.* **48,** 30–55.

Slepyan, E. I. (1961). [Comparison of galls and terates caused by insects with fruits and seeds.] *Bot. Zh.* **46,** 1702–1717.

Smith, D. C. (1979). From extracellular to intracellular: the establishment of a symbiosis. *Proc. R. Soc. Lond.* B **204,** 115–130.

Sonea, S. and Panisett, M. (1984). "A New Bacteriology." Jones and Bartlett, Boston.

Southwood, T. R. E. (1985). Interactions of plants and animals: patterns and processes. *Oikos* **44,** 5–11.

Stone, A. R. (1979). Coevolution of nematodes and plants. *Symb. bot. upsal.* **22**(4), 46–61.

Taylor, F. J. R. (1983). Some eco-evolutionary aspects of intracellular symbiosis. *Int. Rev. Cytol., Suppl.* **14,** 1–28.

Tempé, J., Petit, A., and Ferrand, S. K. (1984). Induction of cell proliferation by *Agrobacterium tumefaciens* and *A. rhizogenes*: a parasite's point of view. In "Genes Involved in Microbe–Plant Interactions" (D. P. S. Verma and Th. Hohn, eds), pp. 271–286. Springer, Vienna and New York.

Thompson, J. N. (1986). Patterns in coevolution. In "Coevolution and Systematics" (A. R. Stone and D. L. Hawksworth, eds), pp. 119–143. Clarendon Press, Oxford.

Trench, R. K. (1980). Integrative mechanisms in mutualistic symbioses. In "Cellular Interactions in Symbiosis and Parasitism" (C. B. Cook, P. W. Pappas, and E. C. Rudolph, eds), pp. 275–297. Ohio State University Press, Columbus.

Tudzynski, P., Rogmann, P., and Neuhaus, H. (1986). Extrakaryotic inheritance: mitochondrial genetics. *Progr. Bot.* **48,** 249–259.

Vaeck, M., Reynaerts, A., Höfte, H., Jansens, S., De Beuckeleer, M., Dean, C., Zabeau, M., Van Montagu, M., and Leemans, J. (1987). Transgenic plants protected from insect attack. *Nature, Lond.* **328,** 33–37.

van Steenis, C. G. G. J. (1969). Plant speciation in Malesia, with special reference to the history of non-adaptive saltatory evolution. *Biol. J. Linn. Soc.* **1,** 97–133.

Van Valen, L. M. (1983). How pervasive is coevolution? In "Coevolution" (M. H. Nitecki, ed), pp. 1–19. University of Chicago Press, Chicago and London.

Verma, D. P. S. and Nadler, K. (1984). Legume-*Rhizobium* symbiosis: host's point of view. In "Genes Involved in Microbe–Plant Interactions" (D. P. S. Verma and T. Hohn, eds), pp. 57–93. Springer, Vienna and New York.

Wade, M. J. and Stevens, L. (1985). Microorganism mediated reproductive isolation in four beetles (genus *Tribolium*). *Science, N.Y.* **227,** 527–528.

Wallin, I. E. (1927). "Symbioticism and the Origin of Species." Williams and Wilkins, Baltimore.

Weis, A. E. (1982). Use of symbiotic fungus by gall maker, *Asteromyia carbonifera* to inhibit attack by the parasitoid, *Torymus capile*. *Ecology* **63,** 1602–1605.

Went, F. W. (1971). Parallel evolution. *Taxon* **20,** 197–226.

Werren, J. H., Skinner, S. W., and Huger, A. M. (1986). Male-killing bacteria in a parasitic wasp. *Science, N.Y.* **231,** 990–992.

Wertheim, M. (1986). Parasites provide a motive for sex. *New Scientist* **109**(1496), 22.

White, P. R. (1951). Neoplastic growth in plants. *Q. Rev. Biol.* **26,** 1–16.

Williams, P. H., Aist, J. R., and Bhattacharya, P. K. (1973). Host-parasite relations in cabbage clubroot. In "Fungal Pathogenicity and the Plant's Response" (R. J. W. Byrde and C. V. Cutting, eds), pp. 141–155. Academic Press, London, and New York.

Wulff, J. L. (1985). Clonal organisms and the evolution of mutualism. In "Population Biology and Evolution of Clonal Organisms" (J. B. C. Jackson, L. W. Buss, and R. E. Cook, eds), pp. 437–466. Yale University Press, New Haven, Conn.

Zambryski, P., Goodman, H. M., van Montagu, M., and Schell, J. (1983). *Agrobacterium* tumor induction. In "Mobile Genetic Elements" (J. A. Shapiro, ed), pp. 505–535. Academic Press, New York.

Index